WITHDRAWAL

Savages and Beasts

Animals, History, Culture

Harriet Ritvo, Series Editor

SAVAGES AND BEASTS

The Birth of the Modern Zoo

Nigel Rothfels

و

The Johns Hopkins University Press

Baltimore & London

© 2002 Nigel Rothfels
Printed in the United States of America on acid-free paper
9 8 7 6 5 4 3 2 1

The Johns Hopkins University Press
2715 North Charles Street
Baltimore, Maryland 21218-4363
www.press.jhu.edu

Library of Congress Cataloging-in-Publication Data
Rothfels, Nigel.
    Savages and beasts : the birth of the modern zoo / Nigel Rothfels.
        p. cm. — (Animals, history, culture)
    Includes bibliographical references and index.
    ISBN 0-8018-6910-2 (hardcover : alk. paper)
    1. Zoos—Germany—History. 2. Wild animal trade—Germany—
History. 3. Hagenbeck, Carl. I. Title. II. Series.
    QL76.5.G3 R68 2002
    590′.7′343—dc21

                                                    2001005689

A catalog record for this book is available from the British Library.

To my parents

Contents
❧

Illustrations

࿋

I am especially indebted to Friedrich and Irmgard Andrae of Hamburg/Volksdorf, who provided me with a loving and welcoming home in Germany when I needed it desperately, and to Donald Fleming, who, as a dissertation advisor and as a friend, has shepherded this project along from its very earliest beginnings. Without their simply vital encouragement and support, not even the research for this book would have ever begun. I wish to express my gratitude as well to Robert Allen Goldberg of the University of Utah for making me want to be a historian and to Kathleen Woodward, former director of the Center for Twentieth Century Studies at the University of Wisconsin–Milwaukee, and Carol Tennessen, the center's executive director, for providing me with an ideal environment to frame my ideas for this book.

For their advice and support at crucial moments in this long process, I am grateful to Rory Browne, Charles Maier, and David Blackbourn of Harvard University; Simon Schama of Columbia University; Elizabeth Frank of the Milwaukee County Zoo; Herman Reichenbach of Hamburg, Germany; Steve Johnson of the Wildlife Conservation Society/Bronx Zoo; Michele Rubin of Writers House, Inc.; Garry Marvin of the University of Surrey Roehampton; Thomas Dunlap of Belmont, Massachusetts; Grace Buonocore of Southbury, Connecticut; Andrew Isenberg and William Chester Jordan of Princeton University; Werner Kourist of Linz am Rhein; Tyler Horsley of New York City; John Edwards of London; Anthony Peressini of Marquette University; Theresa Mangum of the University of Iowa; Lydia Spitzer of Galaxy Hill; Richard Reynolds and Fred Pfening Jr. of the Circus Historical Society; Fred Dahlinger and the remarkable staff of the Circus World Museum; and Carla Bagnoli, Jason Brame, Marcus Bullock, Michael Dintenfass, Lane Hall, Patrice Petro, Glen Powell, Ronald

Ross, Patti Sander, Sylvia Schafer, and Edward Wesp of the University of Wisconsin–Milwaukee. I would like to extend, as well, my special appreciation to the descendants of Christoph and Elisabeth Schulz, including their daughter-in-law Ursula Schulz and their grandson Jurgen Schulz and his wife, Jana Schulz, for their generous advice and assistance. Finally, for substantial material and other support for this project, I am particularly thankful to the German Academic Exchange Service (DAAD); the Minde de Gunzburg Center for European Studies at Harvard University; the Shelby Cullom Davis Center for Historical Research at Princeton University; Dr. Claus Hagenbeck and Caroline and Joachim Friedrich Weinlig-Hagenbeck of Hagenbecks Tierpark; and the Center for Twentieth Century Studies at the University of Wisconsin–Milwaukee.

This book is dedicated to my parents, John and Kathy Rothfels, who always helped me find my own way; it was written because of the love and support of my partner in all things, Heather Hathaway.

Savages and Beasts

In a slim book entitled *Beobachtungen über die Psyche der Menschenaffen* (Observations on the psyche of the great apes), published in 1908, the German zoologist Alexander Sokolowsky attempted to describe the psychological, intellectual, and emotional lives of gorillas, chimpanzees, and orangutans. Convinced that the different anthropoid species had distinct psychological constitutions, the scientist was especially interested in explaining why gorillas seemed to die within days of their arrival in Europe. Noting that they showed a "total lack of engagement with their surroundings," he argued that "one notices immediately that the animals cannot get over the loss of their freedom."[1] Sokolowsky writes: "They sat quietly in their den, without concerning themselves with anyone, and at the most toyed with a blade of straw. Their spirits became more and more gloomy; I could even observe how, in order not to be seen, they tried to keep the eyes of the viewers away by covering their own eyes with their hands."[2] Eventually, the gorillas would be found dead, usually lying face down. For Sokolowsky, the cause of the animals' apparently inevitable demise in captivity was obvious. "It is absolutely clear, from the aggregate behavior of the animals," he argued, "that before all else spiritual influences undermine the health of captive gorillas." Maintaining that the "life energy enjoyed by the animals in complete freedom is sufficient to overcome the dangers posed by parasites," Sokolowsky concluded that in captivity "this energy is broken, the animals give in to their fate, refuse food, and waste away."[3] In the end, the gorillas died of a deep sadness, even melancholy, stemming from their tragic realization of their destiny.[4]

Sokolowsky's study of the psychological life of primates points to larger historical questions of how Western cultures have sought to understand exotic animal life. Studying gorillas in Europe in the first decade of the twentieth century, Sokolowsky saw pathetic, retiring creatures and seemed to find in them a certain thoughtful, or what he called "philosophical," presence. Arguing that the gorilla "thinks more slowly, calmly, and clearly" than the other apes, he concluded with only a little reservation, "I would grant the gorilla a higher spiritual or intellectual gift, because in the final analysis a more elevated development must lie in cautious actions and deeper thoughts than in quick, but also fleeting and restless activity."[5] Sokolowsky's sentimentality, his romantic associations of intellect with loneliness, and his anxieties about reckless energy mark his work as stemming from a particular historical moment.

Indeed, the distinctiveness of that moment is clear in the contrast of his description of gorillas with both earlier and later accounts of the animal. In one of the earliest accounts of the beast, for example, Paul Du Chaillu describes hunting a creature in 1856 which had little in common with Sokolowsky's sad philosophers. After a long and careful stalk, Du Chaillu suddenly found himself confronted by a gorilla standing erect, almost six feet high, with an "immense body, huge chest, and great muscular arms, with fiercely-glaring large deep gray eyes, and a hellish expression of face, which seemed . . . like some nightmare vision."[6] The gorilla, according to Du Chaillu, approached slowly, roaring in rage.

> His eyes began to flash fiercer fire as we stood motionless on the defensive, and the crest of short hair which stands on his forehead began to twitch rapidly up and down, while his powerful fangs were shown as he again sent forth a thunderous roar. And now truly he reminded me of nothing but some hellish dream creature—being of that hideous order, half-man half-beast, which we find pictured by old artists in some representations of the infernal regions. He advanced a few steps—then stopped to utter that hideous roar again—advanced again, and finally stopped when at a distance of about six yards from us. And here, just as he began another of his roars, beating his breast in rage, we fired, and killed him.
>
> With a groan which had something terribly human in it, and yet was full of brutishness, he fell forward on his face. The body shook convulsively for a few minutes, the limbs moved about in a struggling way, and then all was quiet—death had done its work, and I had leisure to examine the huge body.[7]

The illustration in Du Chaillu's volume accompanying this decisive confrontation (fig. 1) shows a somewhat dwarfed white hunter, gun bearer at his side, preparing to fire at a chest-beating monster standing fully upright with

Fig. 1. Paul Du Chaillu, "Death of the Gorilla." *Explorations and Adventures in Equatorial Africa; with accounts of the manners and customs of the people, and of the chace of the Gorilla, Crocodile, Leopard, Elephant, Hippopotamus, and other animals* (New York: Harper, 1861).

somewhat bowed legs. While this illustration carefully coincides with Du Chaillu's written description of the event, the frontispiece to the book (fig. 2) depicts the nemesis in what would become its classic pose, a pose to be echoed in museums around the world for more than a century. The nightmare creature of Du Chaillu's frontispiece, the ultimate threat to the seemingly frail, white, male hunter who is able to conquer the beast only by virtue of his superior intelligence and his courage—if also his gun—illustrates well the essential features of this particular gorilla fantasy: the massive and powerful arms, hands, and feet, the deadly canines, sloping head and protruding brow, the powerful pectoral muscles, the hairiness, and, of course, that something that we are not supposed to see hidden by a judiciously placed branch.

Just how different the representations of gorillas have been over the past century and a half is clear in the contrast between this frontispiece of Du Chaillu's *Explorations and Adventures in Equatorial Africa* and the cover of Sokolowsky's volume. Whereas Du Chaillu in 1861 shows us a menacing creature apparently ready to rip his civilized human cousin limb from limb, the small photograph on Sokolowsky's cover of 1908 (fig. 3) shows us a very young

Fig. 2. Paul Du Chaillu, frontispiece. *Explorations and Adventures in Equatorial Africa; with accounts of the manners and customs of the people, and of the chace of the Gorilla, Crocodile, Leopard, Elephant, Hippopotamus, and other animals* (New York: Harper, 1861).

gorilla sitting hunched forward with the frayed ends of a rope dangling from its neck. Referring to the photo of this animal, which lived for a few days in the Hamburg shop of the exotic animal dealer August Fockelmann, Sokolowsky pointed to the animal's "unmistakable" expression of "anxiety and helplessness" in its "worriedly introspective eyes." Just how "unmistakable" the expression on the infant gorilla is seems open to debate; what is not, however, is Sokolowsky's deep sense of empathy with the animal, an empathy alien to Du Chaillu.

Unnatural Histories

The examples of Sokolowsky and Du Chaillu demonstrate that, while individual gorillas have unique identities, our access to those identities is constrained by the mediated nature of their presence in our historical record. As far as the historical record is concerned, gorillas of the past do not represent themselves; rather, Du Chaillu's "gorilla," Sokolowsky's "gorilla," and even Dian Fossey's "gorilla" are entities inextricably bound by particular human contexts and human interpretations. That is, there is an inescapable difference between what an animal *is* and what people *think* an animal is. In the end, an animal or species is as much a constellation of ideas (for example, vicious, noble, intelligent, cruel, caring, brave) as anything else. And, as with history itself, each generation seems to remake its animals. But even if animals are always only what we think they are, we should not be deterred from exploring their historical record. Indeed, this book began when I started to wonder about how we could talk about "wild" animals in historical ways that made clear the human contexts in which animals have been seen, studied, collected, or killed—in ways that accepted from the outset that animals have been, are, and will likely continue to be understood primarily within human historical settings.[8]

Early on in this project, I came across a passage in Theodore Roosevelt's 1886 *Hunting Trips of a Ranchman: Sketches of Sport on the Northern Cattle Plains* which has stayed with me throughout: "Thus, though gone the traces of the buffalo are still thick over the land. Their dried dung is found everywhere, and is in many places the only fuel afforded by the plains; their skulls,

Fig. 3. Alexander Sokolowsky, cover photograph of gorilla. *Beobachtungen über die Psyche der Menschenaffen* (Frankfurt: Neuer Frankfurter Verlag, 1908).

which last longer than any other part of the animal, are among the most familiar of objects to the plainsman; their bones are in many districts so plentiful that it has become a regular industry, followed by hundreds of men (christened 'bone hunters' by the frontiersmen), to go out with wagons and collect them in great numbers for the sake of the phosphates they yield; and Bad Lands, plateaus, and prairies alike, are cut up in all directions by the deep ruts which were formerly buffalo trails."[9] During a time in which Western cultures were encountering wild, exotic animals in more startling numbers and varieties than perhaps ever before or since, Roosevelt pondered what had once existed while mulling over that which had been left behind. His words suggest the historian's task: piecing together an earlier time by somehow bringing back to life the muted remnants left behind. We do not have to look far today to see that the traces of exotic animals are "still thick over the land": they lie in yellowing photographic albums; they slowly rot and burst their seams in museums of natural history; they are old, warped, and cracked billiard balls and ancient, elephant-foot ashtrays; they are the ivory-handled heirloom cutlery sets still brought out on special occasions and the trophies of great hunts hanging in museums, sports bars, and living rooms. The animals of the past inhabit musty books and paintings, faded journals and memories, but their traces are more than just the relics of *their* existence; their remains are also the people who were fascinated by them, who studied them, who hoped to learn from them, and who were sometimes preoccupied with catching or killing them.

I have often thought about Roosevelt's response to the buffalo trails when I've leafed through old works of natural history like that of Du Chaillu and Sokolowsky or when I've walked through museums of natural history and zoos. I've thought about the literally millions of animal remains in American and European museums and realized that almost every one of them must have a small tag upon which one could read where, when, and by whom the animal was collected. Every time I have begun to consider these animal fragments, I've come to the conclusion that these animals' *un*natural histories—their lives and afterlives not in their "native haunts" but in such human environments as museums, books, circuses, and zoos—are at least as worth telling as their natural histories. I have also found that these unnatural histories are usually a good deal more interesting—at least to me—than knowing the taxon with which the animals have been associated by zoologists. Finally, I've found that these unnatural histories are vitally important, as is suggested by our seemingly collective attempt as humans to try so hard not to see them.

Visiting the spectacular exhibits at the American Museum of Natural History in New York, for example, I see a moment in a northern forest when two massive bull moose engage in a titanic battle. Just around the corner, I peer into a dark, moonlit winter landscape to see two wolves effortlessly leap into the woods from a clearing. Soon I find myself scanning a Wyoming vista in which a herd of buffalo stretches into the distance, while in the foreground I examine the huge animals in detail in their prairie environment. But this is not a prairie, of course, the wolves are not in the woods, the moose are not in the midst of battle. All these animals are in a museum in Manhattan; they were all shot long ago by someone who skinned them and preserved their hides, so why is it (or how is it), then, that we can never see where the bullet entered the beast? Beyond all that, someone had collected plants and taken photographs of the scene. Someone painted a backdrop, and someone stuffed the animals so that they would look "lifelike." Why is it important to us that we hide real elements in the unnatural histories of these animals so that we can see their natural history in such detail?

Visiting a zoo presents a similar set of problems. We all know, for example, that at their most basic level zoos are for people and not for animals. But visiting one of the premier zoological gardens of today, we quickly find ourselves immersed in environments designed to mask the fundamentally and overwhelmingly human nature of the place; instead of seeing animals in the ornate buildings of the late-nineteenth-century zoological garden, we now see them—if often only with difficulty—in deeply wooded "forests" and "jungles" or at a distance as they move slowly across a stretching "veldt." There are issues of good husbandry behind modern zoo exhibits, and certainly there is an educational value in "enclosures" (no longer called cages) that try to simulate an animal's more natural environment. But the primary motivation behind these exhibits is unarguably aesthetic: we don't like to see "wild" animals pacing behind bars and being poked and gawked at by rude visitors. Indeed, aesthetics are deeply embedded in the zoo experience. At my local zoo when I was growing up in Utah in the late sixties and early seventies, for example, there was a liger (the hybrid offspring of a lion and a tiger) whose name was Shasta. I, like most people, was fascinated by Shasta's beauty, even if, with the crowds of people, it was often something of a struggle to get a good look at her. When she died in 1972, and because she was locally famous, she was stuffed and put back in the cat house in a special glass case. People still came to the zoo to see Shasta. Now, thirty years later in a climate in which zoos need more than ever to justify their existence as conservation organizations, however, it is perhaps not too surprising that Shasta is no longer to be seen at the zoo. She doesn't fit the current aesthetics. We don't *like* to see stuffed animals at the zoo—the zoo, it is argued, should be understood as a place that attempts to re-create the natural habitat of animals in a world in which the dangers are too great for animals actually living in their natural habitats. The zoo, by this line of reasoning, is not a place for the entertainment of people but for the preservation of animals.

Carl Hagenbeck

My interests in the unnatural histories of animals led me to this study of the hunting, capturing, trading, and exhibiting not only of animals but also—and, at first, surprisingly—of people in the late nineteenth and early twentieth centuries by the company of Carl Hagenbeck in Hamburg, Germany. An older generation of Americans might remember the name Hagenbeck from the Hagenbeck-Wallace Circus, but few people outside Germany who have not explored the history of zoos will know it. For most contemporary Germans, however, "Hagenbeck" still means a zoo outside Hamburg and a family that has been involved with the spectacularly exotic for as long as anyone seems to be able to remember. For people interested in the history of zoos, the name has come to mean something more specific. It refers to what has come to be known as the "Hagenbeck revolution" at the beginning of the twentieth century—generally seen as the breakthrough moment of modern zoo design when animals were freed from their cages to live in naturalistic enclosures. But while the "Hagenbeck revolution" points to a specific place at the beginning of the twentieth century, the foundation of the company traces back to the mid-nineteenth century, when Hagenbeck's father, a fishmonger in Hamburg, began buying and selling exotic animals arriving at the port city. By the 1870s—and here the fortunes of the company paralleled those of similar firms in Hamburg concentrating on such other natural resources as guano, sugar, coffee, palm and whale oil, or rice—the business had evolved from a sideline interest of a small fish shop to the world leader in the international trade in exotic animals, a position that remained unchallenged until the beginning of World War I. Practically every major zoological garden, circus, or private collector anywhere bought animals from Hagenbeck.

Carl Hagenbeck did not restrict his business simply to trading exotic animals, however. Perhaps most striking as we look back was the company's decision in 1874 to begin procuring indigenous people from all over the world for presentation in highly profitable spectacles to European scientific societies and the general public. While continuing his profitable trade in animals and people, Hagenbeck then began in the late 1880s to exhibit a series of unique animal acts, the animals for which, he claimed, had been trained in altogether new and humane ways. Finally, in 1907, the firm's animal business, exhibitions of people, and performing animal acts found a permanent home in Hagenbeck's new Animal Park (fig. 4), a zoo without the iron bars that had become the most discomforting object for visitors to the older zoological gardens. Based on experiments begun more than a decade before, Hagenbeck's park, with its panoramas of animals in which the animals were separated from one another and the public through carefully hidden moats, became the model for the remainder of the twentieth century. Here, animals appeared to be living in the wilds of Africa or India even though they actually lived in a zoo in northern Germany. At Hagenbeck's park visitors could observe "exotic" animals and even peoples in their "native habitats"—the African jungles, Russian steppes, American plains, and Arctic ice—without ever encountering a bar or visible barrier and without ever leaving the comfort of their own "civilization."

By telling the story of Carl Hagenbeck and his unusual firm, and by recording as well the unnatural histories of the animals and people who became the exhibited objects at Hagenbeck's, this book contributes to a growing body of studies examining the origins and development of the trade in exotic animals and peoples. At the same time, it extends our understanding of the history of zoological gardens, as well as the history of the rapid international growth of German businesses in the second half of the nineteenth century and the effectiveness with which these businesses often utilized the resources of new colonial acquisitions. Perhaps most important, however, this book shows how Hagenbeck's various enterprises influenced popular representations of exotic cultures and wildlife in the crucial years before 1914 and how the legacy of those representations lives on to this day. Beyond all this, though, I hope this book provides readers with the historical background they need to grapple

Fig. 4. Opening Day at Hagenbeck's Animal Park, May 1907. *Von Tieren und Menschen: Erlebnisse und Erfahrungen* (Leipzig: Paul List, 1908). Courtesy of Hagenbeck's Tierpark.

with the vexing issues that zoological gardens present to us today. Because Hagenbeck's various interests are deeply embedded within the goals of the modern zoological garden, understanding the importance of his company should bring into clearer focus the basic problems inherent in the modern zoo project—a project that must be traced directly to Carl Hagenbeck's exhibitions of animals and people.

At the core of this book is a common experience we have all had of looking at exotic animals at the zoo, at the circus, or at the pet shop. For most of us, most of the time, there is something usually unsatisfactory about this experience. Pointing to the constant questions of children about why the animals at the zoo don't move, don't do anything, don't seem to care about anything, the art and cultural critic John Berger, for example, summarized the disquiet in a simple question: "Why are these animals less than I believed?"[10] Berger's answer begins with what he calls the marginalization of animals in the industrializing nineteenth century—their removal from the centrality of human experience. He writes, "Public zoos came into existence at the beginning of the period which was to see the disappearance of animals from daily life. The zoo to which people go to meet animals, to observe them, to see them, is, in fact, a monument to the impossibility of such encounters."[11] Fundamental to the disappointment of the zoo, Berger notes, is that the animals seem to resist looking at the visitors. This point becomes more clear when we consider that museums of natural history almost always attempt to deliver precisely what zoos cannot. With total control over the displayed object, the exhibited animal or human at the museum of natural history is forever caught unaware by the museum visitor. Seemingly invisible, the museumgoer steps lightly beside the water hole or walks through the "native village," and the exhibited animals and people both seem and do not seem to notice. On the one hand, unlike at the zoo, the objects in the natural history museum's display cases are not practiced in looking away from the visitor and most often, in fact, look out directly to the viewers, thus appearing somehow to be aware of a new presence. On the other hand, the animals and people exhibited at the museum are—again highlighting the problems of zoos—nevertheless always actively engaged in their own lives and somehow seem oblivious to the strange environs in which they have been placed.

The point is that even with Hagenbeck's revolutionary ideas of zoo design—ideas that have been expanded and elaborated upon throughout the last century by others who continue to believe that the ideal exhibit will soon be created—the zoo still ultimately disappoints. Just how disappointing it can be becomes clear through a few questions: How many times over the years have we witnessed someone at the zoo trying desperately, obnoxiously, violently to get the attention of an animal? How many stones, sticks, coins, soft drinks, and all manner of other items have been thrown into cages to try to get a response from an animal? Why do throngs descend when it is time to feed the lions or the penguins? Why are people willing to pay extraordinary sums for the opportunity to swim with dolphins in a small tank? Why is it that when the panda moves from one spot to another at the National Zoo in Washington, D.C., or when the handlers enter the elephant enclosure to work with the animals, or when a parent says too loudly to a child, "Look, that gorilla is looking at *you*," crowds seem to gather instantly? We seem intent to catch the look of an animal, to see the animal look at us. This is, of course, the look that so many hunters talk about, the look caught in taxidermy the world over, a look described beautifully in the opening page of Richard Nelson's meditation on the meaning of deer in America, *Heart and Blood:* "The buck stands atop a rise, darkly silhouetted against broad-trunked oaks and maples strong with garlands of freshly opened leaves. He watches nervously but holds his ground, as if my stillness has assuaged his fears. . . . At last he turns casually away, leaning down to nibble tender shoots of grass, and when he looks up again a tuft of silvery dandelion fuzz clings to his chin like a beard. Every few seconds—as if he's twinged by a dissipating memory—he abruptly raises his head and stares at me, focusing his outsized ears for the slightest betraying sound."[12]

Much of what follows in this study of looking at the exotic at the end of the nineteenth century turns, in fact, on awkward and uncomfortable moments at zoological gardens when the exhibited objects either look back or do not look back at the viewer. I hope it becomes clear that regardless of how much we may desire the animals in our collections to return our look, regardless of how much we are disappointed when they try to escape our attention, it is precisely when they *do* look back that the edifice undergirding the zoological garden begins to collapse. In the end, little is more destabilizing to our vision of the zoological garden than when the animal looks back at us and seems to know something about both us and its own predicament.

Certainly, this was the experience of the shows of exotic people at zoological gardens. The shows worked only so long as members of the viewing public could convince themselves that either (1) the exhibited people did not really understand what was going on or (2) the exhibited people were better off in a zoo in Germany than leading a "miserable" existence in some remote and isolated camp halfway round the world. Once the exhibited people in the shows began consistently to refuse to be seen as "fabulous animals," once they began to look back and interact with the viewing public as fellow humans, the people shows were doomed. As the people in the shows began to talk back, however, the animals were carefully "silenced," and this enforced "silence" is perhaps the defining feature of the modern zoo. If one or two animals occasionally act out against it—when, for example, a heretofore quiet gorilla or always endearing chimp viciously attacks one of its keepers and thereby demonstrates to the public that his or her life is not always as we might like to imagine—the animal is quickly silenced again, sometimes for good. Meanwhile we continue to bring our children—and to go ourselves—to the wilds of the zoo, a place that seems to fascinate more than any other.

Centuries before the firm of Carl Hagenbeck came to dominate the trade and representation of exotic animals, unusual animals had been exhibited in major cities around the world. In order to understand the importance of Hagenbeck and his innovations, we must first look back at the longer history of those exhibitions. To put the point as simply as possible, it seems that wherever and whenever there have been cities, there have been collections of unusual animals. Indeed, going back deeply into most human civilizations, we can trace these kinds of collections, though we know most, of course, about large collections assembled by political figures in major cities. In the ancient world there were large Babylonian, Chinese, and Greek collections, and exotic animals were kept by all sorts of Roman emperors, governors, and statesmen, including Alexander the Great and the Emperors Trajan (who, it is reported, had a collection of some eleven thousand animals) and Nero. At the end of the eighth century, Charlemagne owned a large menagerie, and in the twelfth and thirteenth centuries there were large collections in France, Italy, the German-speaking countries, and England. (Henry I, son of William the Conqueror, for example, started an English royal tradition of keeping exotic animals when he built up a collection at Woodstock. The collection was later moved to the Tower of London by his grandson, Henry III, and only eventually dissolved in the mid-nineteenth century when the rather poor remnants were transferred to the new Zoological Gardens of London in Regent's Park.) In the sixteenth century, we have records of both the thousands of animals belonging to Akbar, the third Mogul emperor of India, and Montezuma's huge collection in Tenochtitlán. Throughout these thousands of years, moreover, smaller collections, sometimes consisting only of indigenous, nondomestic animals,

Fig. 5. Jean-Baptiste Oudry, *Study of a Rhinoceros*, 1749. © Copyright The British Museum.

were maintained by all manner of individuals and groups. In Europe, for example, the dry moats surrounding walled cities were often stocked with deer, songbirds were kept in elaborate aviaries, falcons and cheetahs were kept for hunting, and bear collections were relatively commonplace. On top of all of this, of course, traveling menageries and fairs brought unusual animals to cities at regular intervals.

Then there are those single famous animals (including elephants, rhinoceroses, giraffes, and hippopotamuses) that were transported at length across Asia, Europe, and the New World. Among these, the rhinoceros known as Clara, who traveled throughout Europe from 1741 to 1758, is only one of the more famous. During the course of eighteen years, Clara was brought to cities in the Lowlands, the German- and Polish-speaking countries, France, Italy, England, and Sweden.[1] Indeed, it was finally the prodigious artistic output resulting from Clara's travels, including the many posters commissioned by her handler, Douwe Mout, drawings by Pietro Longhi, Johann Elias Ridinger, and Jean-Baptiste Oudry (fig. 5), and objets d'art such as Louis XV ormolu clocks, which finally began to unseat Albrecht Dürer's famous 1515 woodcut representation of the rhinoceros as a sort of armor-plated ramming machine, an image that had been the accepted visual representation of the animal for more than two hundred years.[2] From royals to aristocrats, merchants to commoners, Clara was seen, drawn, studied, and even, as T. H. Clark noted, emulated in a rage of rhinocerotic hairstyles.[3]

That exotic animals have, in fact, been fully a part of our cultural language for hundreds of years is clear in Pieter Bruegel the Elder's small work from 1562 depicting two chained monkeys (fig. 6). In the foreground of the piece, the monkeys sit in a thick, arched, stone viewing hole overlooking a harbor. Around their waists are metal rings attached to massive metal chains, themselves attached to a further metal ring centrally affixed in the window's sill. One of the monkeys faces us, looking out of the picture engagingly; he is both teacher and apprentice, and with Bruegel's name below him, perhaps he sits in for artist himself. His future of captivity is the fate of both monkeys, but rather than sinking into depression and remorse, he watches, fascinated by the events around him. A captive animal, he studies the nature of captivity. The other monkey, facing away from us, sits crumpled and depressed, brooding over his

Fig. 6. Pieter Bruegel the Elder, *Two Chained Monkeys,* 1562. Berlin, Gemäldegalerie, Staatliche Museen.

view out over water, ships, birds gliding high in the air, and the blue-gray skyline of a city beyond. Like an old beggar, he seems to avert our gaze and looks down, alternatingly gaping vacuously at the shards of hazelnut shell at his feet and glimpsing the open world beyond, trying desperately to see the connection between the discarded shell and his fate. He is broken down, bitter. Bruegel's moral allegory has been repeatedly explained by historians of art: the monkeys were caught in their haste to eat the nuts set in a trap—the lures of easily gotten gains and ephemeral pleasures seduced the animals into dropping their guard, and though the nuts were won, the monkeys lost their freedom and innocent happiness. But if the painting is a parable, Bruegel's more political message is also evident: the skyline of Antwerp seen out the monkeys' window points to the political, commercial, and religious crises of the city, Flanders, and the Lowlands more generally which were consequent to the Spanish king Phillip II's succession in 1557 and his declaration of bankruptcy in 1559. Long one of the "economic pillars" of the Habsburgs, the fortunes of Antwerp were so closely tied to the growth and solvency of the Empire that the Empire's financial demise spelled an almost inevitable commercial and governmental catastrophe for the city.[4]

Beyond all moral and political messages, however, what must be emphasized is that Bruegel's painting of the chained monkeys made sense in the sixteenth century because he and his viewers had seen monkeys—mangabeys—in captivity and had contemplated the issues of captivity which the animals' circumstances posed. Although Bruegel's specific critique might have been about the changing political climate of Europe, he uses the vocabulary of exotic animals in captivity—a vocabulary with which his audience is thoroughly familiar—to make his point. There remains, however, a significant difference between pointing out that there were a good many strange animals to be seen in Europe over the centuries and trying to figure out what the significance of those animals was to their viewers. This is the point made in the differing accounts of gorillas coming from Alexander Sokolowsky and Paul Du Chaillu with which this book began. Indeed, the history of trying to figure out that significance is worth exploring here briefly.

Perhaps not surprisingly, for a period of rapid economic and social change, increasing confidence in the future, and an expanding popular interest in studying the past, the first books and articles about the history of the exhibition of exotic species began to appear in the second half of the nineteenth century and the early twentieth. Written by historically inclined members of the burgeoning zoological societies connected with the major zoological gardens of Europe and the United States, these works seemed to have had two primary interests. On the one hand, a significant part of these books was dedicated to the task of documenting as simply and completely as possible both the many different kinds of collections which had been put together over the course of the centuries and the many different kinds of animals shown in those collections. On the other hand, written as they were by supporters of the grand new public gardens, these works tended to adopt an optimistic and progressivist view of animal collections which equated the history of zoos with a more general history of human enlightenment. Thus, in his history of the zoological collections written in German around 1911, for example, Friedrich Knauer states with apparently little doubt, "The times in which one saw little more in zoological gardens than a menagerie of larger proportions, satisfying only the curiosity of the visitors, are in fact over for a large portion of the population. Today, no one who wants to be taken seriously will doubt the educational and scientific roles of a zoological garden, which on the one hand should foster the informing of the public, the advancement of a sense of nature, and the interest in observing animals especially among the youth, and which on the other hand will stand in the service of science."[5] For Knauer—and his international counterparts writing in French and English—the latter half of the nineteenth century witnessed a complete reorientation in the keeping of exotic animals.[6] Arguing that the new zoological gardens had gratifyingly become institutes of learning and science, places where children could be brought and taught to appreciate and value the diversity and even sanctity of nature, he dismisses as ignorant, or perhaps simply not serious enough, those who retained doubts about the importance of the collections.

In his 1911 appeal for continued financial support from the city of Hamburg for its zoological garden, however, Dr. J. Vosseler's ambitions were slightly more grandiose. Arguing that the "impression of the living, moving, and behaving animal is neither replaceable through museum exhibits nor pictures and is only to be gained in a zoological garden," Vosseler concluded, "Intimacy with the living world makes people indigenous, and awakens and sustains the sense of home and the love of Nature and her creatures as the best counterbalance to the social disadvantages of modern life. The garden, therefore, supports the school and the family and thereby the city organism in ethical and moral education, and contributes directly to the efforts of the movements for the protection of animals and nature."[7] The director of the zoological gardens in Frankfurt, David Friedrich Weinland, rounded out this sort of a laudatory chorus in his 1862 article on the origins of zoological gardens, pointing to what he considered to be the most distinguishing characteristic of the modern collections. "Neither princes, nor scholars, nor pedagogues, nor ministers of education founded the zoological gardens of Frankfurt, Dresden, Cologne, Hamburg, Amsterdam, Antwerp, Rotterdam, and Brussels," he argued. "Rather they were created by the majority of the citizens of these cities, who were driven—one almost wants to say—by an unconscious yearning to watch living nature."[8] Simply, according to Weinland, the educated citizens of the flourishing cities had organized and constructed zoological gardens because they provided an attractive, educational, and apparently psychologically necessary diversion for the urban dweller.

The central reasons given by these authors for the existence of zoological gardens—education, science, the possibility that the collections could become centers of emerging environmental and animal protection movements, and a basic human need for nature in the midst of urban concentration—were repeatedly claimed throughout the twentieth century by institutional historians with similar investments in the goals of modern zoos. For example, Heini Hediger, one of the foremost zoo biologists of the second half of the twentieth century, rehearsed these same goals for zoos in 1971: "In contradistinction to earlier times it is not just live curiosities—as it were museum preparations in motion—that are sought in a zoo; the public now feels a growing need of, and has a right to, relaxation and recreation in a zoo—in a zoological garden. There lies and remains the primary function. Added to this are three others, viz. popular information, research and nature conservation also in terms of offering sanctuaries to jeopardized species, and the preservation of their living space."[9] For these historians of zoos, the annals of the collections have evolved into an illustration of the history of the West in which civilization has slowly come to appreciate, value, and conserve nature. This story begins, of course, with the extravagance and waste of the Roman Empire with its bloody displays in public arenas, becomes suitably murky in the "Dark Ages," expands again in the Renaissance, and receives new prestige in the princely collections of the seventeenth and eighteenth centuries. It reaches a high point in the French Revolution and the consequent opening of the first *public* zoo in 1793—the Jardin des Plantes—and its "bourgeois" followers such as the Zoological Gardens of London (1828) and the gardens in Amsterdam (1843), Berlin (1844), and Central Park (1862). Finally, it concludes in our own times with the "new zoos," such as San Diego's Wild Animal Park and Disney's Animal Kingdom, where animals roam apparently free and their human observers, ever sensitive to their own non-naturalness, sit quietly with their binoculars and cameras in electric monorails and safari vehicles while wildlife carries on with being natural.[10]

The big division for all these histories is between something commonly called a "menagerie" and something else called a "zoological garden." The former term is generally used pejoratively to describe collections of captive animals kept "simply" for purposes of display or for the aggrandizement of the owner. Thus, both a sixteenth-century noble's collection and a nineteenth- or twentieth-century traveling or fixed collection of caged animals shown for profit are called menageries. Zoological gardens, on the other hand, are understood to be places that privilege scientific endeavor and public education. In the tellingly titled *Zoo Book: The Evolution of Wildlife Conservation Centers*, which was published in cooperation with the American Zoo and Aquarium Association, Linda Koebner writes, for example:

> The private lives of nobility and kings were breaking down; lands and treasures were redistributed. Eventually many menageries were collected to become one. [The] menagerie from Versailles was taken to the Jardin des Plantes, a botanical garden in Paris.
>
> In 1793, the royal animals that had remained in Versailles were also sent to the Jardin des Plantes. It was decided that they should be a collection of scientific value, to be studied as wonders of nature. The idea of the zoological garden had arrived.[11]

This idea that the Jardin des Plantes and its emulators represent a pivotal moment in the history of keeping exotic animals in captivity seems to be generally accepted by practically all "zoo historians." For example, in the introductory chapter—a sort of "prehistory"—to the 1996 *New Worlds, New Animals: From Menagerie to Zoological Park in the Nineteenth Century*, we read: "Up to the beginning of the nineteenth century, little had changed in Western menageries; nearly all were still the province of the nobility and the wealthy. Animals were caged for human amusement and as symbols of status and power. Exhibits were not systematically organized. With few exceptions, animals were displayed throughout the menagerie or zoo for 'the gratification of curiosity and the underlining of the magnificence and power of their own-ers.'"[12] Perhaps needless to say, a paragraph like this begs more questions than it answers. Are we to believe that through the many centuries leading up to "the nineteenth century" the conditions and symbolic importance of animal collec-tions in "Western menageries" were somehow stable and universal? Can we believe that animals in zoos since the beginning of the nineteenth century have not been caged by a wealthy aristocratic class for the purposes of "human amusement and as symbols of status and power"? Can we imagine a collection of exotic animals which has not been organized according to some sort of sys-tematics? If all the blue animals were put in one corner, is not that also a sys-tem? What the paragraph is really telling us, of course, is that animals in earlier collections were not organized around a systematics that we see today as some-how more logical or more enlightened.

If, however, historians of zoological collections have generally been more interested in an antiquarian task of bringing to light the presence of exotic ani-mals in the past and more convinced by a progressivist view of history, histori-ans of culture have been far more successful in examining the meaning and importance of those animals to their collectors. Typically, these historians have focused more on the relations between certain forms of animal exhibition and their cultural milieu than on proving a case for the general amelioration of con-ditions for animals in captivity. As early as Jacob Burckhardt's 1860 *Civiliza-tion of the Renaissance in Italy*, for example, these historians have returned again and again to examine how the keeping of exotic animals is related to other issues of human society. In his effort to suggest the expanding interest in nat-ural history during the Renaissance, Burckhardt naturally looked to the popu-larity of collections of exotic animals and plants in the period. But while not-ing that collectors sought to keep as many species as possible for purposes of study and simple curiosity, Burckhardt did not fail to recognize the symbolic importance of the collections and their inhabitants. He writes: "The cities and princes were especially anxious to keep live lions, even when the lion was not, as in Florence, the emblem of the State. The lion's den was generally in or near the Government palace, as in Perugia and Florence; in Rome it lay on the slope of the Capitol. The beasts sometimes served as executioners of political judg-ments, and no doubt, apart from this, they kept alive a certain terror in the pop-ular mind. Their condition was also held to be ominous of good or evil. Their fertility especially was considered a sign of public prosperity, and no less a man than Giovanni Villani thought it worth recording that he was present at the delivery of a lioness."[13] Burckhardt further records "human menageries," books on natural history, and extraordinary events such as the reception in Florence in 1459 of Pius II and Galeazzo Maria Sforza, which included pitting against one another in the Piazza della Signoria "bulls, horses, boars, dogs, lions, and a giraffe."[14]

In recent years, historians of Western culture have become increasingly attentive to the presence of animals—and nature at large—in the past, and the rewards of their work have been significant. From studies of pet keeping to whale beachings, investigators of the human past have become interested in examining connections between animals, the natural world, and human societies. In the realm of histories of zoological gardens more specifically, cultural critics and historians have pointed to more oblique motivations behind the gardens than education, science, conservation, or amusement and recreation. Pointing to the parallel roles of bourgeois zoos and their princely predecessors, for example, John Berger pointed out that the great zoos of London, Paris, and Berlin

> brought considerable prestige to the national capitals. The prestige was not so different from that which had accrued to the private royal menageries. These menageries, along with gold plate, architecture, orchestras, players, furnishings, dwarfs, acrobats, uniforms, horses, art and food, had been demonstrations of an emperor's or king's power and wealth. Likewise in the 19th century, public zoos were an endorsement of modern colonial power. The capturing of animals was a symbolic representation of the conquest of all distant and exotic lands. "Explorers" proved their patriotism by sending home a tiger or an elephant. The gift of an exotic animal to the metropolitan zoo became a token in subservient diplomatic relations.[15]

This point is echoed by Keith Thomas in his study of the roots and development of a "modern sensibility" toward animals when he argued that the "royal menagerie [at the Tower in London] symbolized its owner's triumph over the natural world" and that "the zoo became a symbol of colonial conquest as well as of wealth and status."[16] Indeed, the notion of the zoo—especially the nineteenth-century zoo—as an embodiment of colonial ideas has become a repeated focus in recent years. Perhaps the most sustained attention to this theme, however, has been given by Harriet Ritvo in her groundbreaking study of various aspects of the relationship between animals and humans in Victorian

England. Attempting to coordinate the many tendencies and impulses behind English attitudes toward animals in the nineteenth century, Ritvo finds in collections of exotic animals—from exotic caravans traversing the English countryside to the London Zoo, from private menageries to provisioners for the hunt—impulses stemming from the English viewing foreign lands through distinctly imperialist objectives.[17] For Ritvo, the motivations behind British collections of exotic animals lay in an overpowering urge to overpower. Other recent interpretations of the origins of modern zoological gardens consider them not specifically as structures of imperialism but rather as more general products of bourgeois sensibilities and economies. The animal collection of the Jardin des Plantes thus becomes the first arena for the "public man" to acquire and study the world of exotic animals, while the great collections of the nineteenth century in London, Berlin, Philadelphia, Antwerp, and so on become echoes of bourgeois inclinations to construct uplifting temples of knowledge backed by capitalist economies.[18] According to this line of reasoning, the gardens, along with the great museums, were part of a saturating bourgeois ethos—as David Blackbourn argues, "a tribute to bourgeois self-confidence: symbols of the power of science and witnesses to how Progress could edify man while taming the natural order."[19]

In effect, two camps have developed around the issue of the significance of zoological collections. On one side are those who argue that whereas such collections may once have been more or less organized expressions of princely power over people and nature—perhaps the result of some "simple" or "vulgar" curiosity, or at worst, the result of some wretched fascination or pecuniary interest—the gardens of the nineteenth and twentieth centuries represent a genuine interest in the animals expressed in a desire to learn from and about them.[20] On the other side stand those who have sought the persistent presence and development of certain social, political, and economic forces behind all collections of exotic animals. These two approaches, of course, yield rather different stories about the meaning of various exhibits at zoos. Of the two approaches, however, the latter is much more able to make sense of a structure like the bear pit, perhaps the most popular animal enclosure of the nineteenth-century garden. Usually consisting of a deeply excavated circular hole about fifteen to twenty feet in diameter with a high tree trunk sunk in the center, bear pits became a focus for amusement in gardens all over Europe and the United States. At the London Zoo, for example, the Zoological Society operated a

Fig. 7. The bear pit at the London Zoo, ca. 1900. Courtesy of John Edwards.

concession near its pit (which was one of the very first structures in the gardens), so that visitors could purchase buns to feed the animals. The bears would then climb the trunk, often three or four at a time, so that the happy spectators could throw food to them and laugh at their "unwieldy gambols" (fig. 7).[21] Whereas the institutional historians of zoos would tend to discount the significance of the long-lasting bear pits and perhaps suggest that they represented only an unfortunate holdover of an older or less sophisticated view of animals, the cultural historians would tend to see the pits as quintessential to the ethos of the major zoological gardens—in the bear pit, the animal that had long been associated with fear, the woods, and aristocratic hunting privileges was reduced in a controlled, urban, and bourgeois environment to a comic figure asked to perform for ladies, gentlemen, and, perhaps most important, children. This is not exactly a side of bourgeois zoos which their champions have found most seemly, but the bear pit should indeed be seen as an integral part of bourgeois ideas about animals.[22]

Still, if cultural histories have offered more illuminating accounts of the *significance* of animal exhibits, most of these accounts have been limited in size and scope and have generally been part of larger discussions that have had little to do with animals. Although the historian Simon Schama, for example, writes insightfully about the exhibition of European bison in the forests of Poland in his essay "The Royal Beasts of Białowieża," animals are not the focus of his remarkable larger study *Landscape and Memory*.[23] Meanwhile, the institutional histories of zoological gardens continue to be written and broadly accepted. At the gift shops of most major zoos, for example, books are available which tell the history of that zoo in usually rather unsurprising narratives detailing the *progress* and *development* of the institution over the years. Again, the basic structure of these histories demonstrates how older menageries became zoological gardens and, in some cases, how those zoological gardens have evolved again into something now called "conservation centers."

This sort of story has become the dominating narrative claimed by our zoological gardens today. According to this reasoning, animals in at least our major zoos are no longer the freak pets of a decadent nobility, no longer the victims of imperial contest; rather, they are the treasured lucky few. As a recent defender of the London Zoo, Colin Tudge, claims by using once again a progressive model, "A hundred years ago—or even a decade ago in many cases—the life of animals in zoos could best be described in the words of Thomas Hobbes: 'solitary, poor, nasty, brutish, and short.' Now, curators of good zoos can effectively guarantee—barring hurricanes and other Acts of God—to keep animals alive in captivity in most cases for far longer (perhaps several times as long) as they could reasonably expect to live in the wild."[24] Removed from the dangers of living in the rarely "wild," often "war-torn," typically "horribly impoverished" areas of the world to which they are indigenous, animals in today's enlightened zoos, with their veterinarians and antibiotics, can now—to paraphrase Tudge—look forward to long lives; indeed, they can look forward to genetic immortality as cryogenically preserved gametes and tissue samples.

Since so much of the reasoning behind these kinds of arguments has been constructed around the conventional, progressivist, institutional histories of zoological gardens, however, we must examine their central argument more carefully. Are there, in fact, disjunctures or discontinuities between menageries, zoological gardens, and now conservation centers? Or, stated another way, are the purposes and goals of the major zoological gardens at the beginning of the twentieth-first century fundamentally different from their historical predecessors? To begin to answer these questions, it is helpful to compare briefly an eighteenth-century menagerie, that of Prince Eugene of Savoy, with a nineteenth-century zoological garden, in this case the Zoological Gardens of London, the model for bourgeois zoological gardens of the period. The comparison shows significantly more continuity than the generally accepted narratives of the history of zoological collections suggest. More important, the comparison provides a necessary backdrop to the truly pivotal importance of Carl Hagenbeck's innovations in presenting exotic animals to the public. At the same time, however, the comparison highlights generally ignored discontinuities—real differences between zoological collections of the early twentieth century and their historical predecessors—which come into even sharper focus when we explore in detail the exhibitions developed by Carl Hagenbeck and when we turn at the end of this book to the rolls of our zoological gardens—or conservation centers—today.

The Belvedere Menagerie

In his 1730 account on the Holy Roman imperial estates, Johann Küchelbecker briefly describes the Belvedere Palace of the great military leader Prince Eugene of Savoy.[25] Belvedere, Eugene's second Vienna residence, stood with a commanding view northeast to the city just beyond the undeveloped military zone surrounding the city's walls. The palace complex consisted of two main buildings separated by a long, sloping, formal garden. Between 1714 and 1716 the architect Johann von Hildebrandt completed the first building, the Lower Belvedere, a relatively unimposing building at the northern end of the grounds with an unbroken south-facing row of windows providing natural light for the winter gardens that framed the living quarters. This delight in plants, and more broadly in nature, was extended in the second phase of construction, between 1716 and 1719, with the development of the estate's formal gardens and menagerie. To contribute to this phase Eugene acquired from the court of Max

Emmanuel of Bavaria the landscape gardener Girard, a pupil of the famed Le Nôtre, the designer of the gardens at Versailles, and his completed gardens echo Louis XIV's huge productions. Finally, from 1720 to 1724, the third phase of construction, focusing on the Upper Belvedere at the top of the slope, was completed. Visitors, arriving at the huge Belvedere and clearing the elaborate main gate, would have their first unobstructed view of the strikingly white, copper-roofed Upper Belvedere while standing before a shallow reservoir reflecting the building. Flanking this stunning head-on view were rows of chestnut trees, which further tended to direct attention to the building and its apparent, only inverted, watery twin.

After a brief look around the two buildings and the rest of the estate, one would quickly realize that Eugene, like so many of his auspicious contemporaries, was an avid collector. His large collection of paintings was varied in style and theme, his collections of scientific instruments and objets d'art were notable in size and scope, and his collection of books, which was complemented by the library in his city palace, was one of the more remarkable of the time. Besides these, however, were two other and, for Eugene, particularly significant collections: his animals and plants. Indeed, of the nine and a half pages of text describing the appointments of the Belvedere, Küchelbecker fills nearly three with a description and partial catalogue of Eugene's menagerie. For the chronicler, the listing of the animals was an important part of demonstrating the wealth and brilliance of Eugene's properties. His more general description of the palace can, in fact, be read almost as an accountant's ledger: paintings in miniature @ 200,000 florins; a painting of a "female embracing a youth in a bath" @ 30,000 florins; a crystal chandelier @ 18,000 florins;[26] or less precisely, but perhaps equally telling, in one room, "The walls are covered with mirrors, and, like the ceiling, heavily gilded, the floors laid with expensive woods of a great variety of colors."[27] To be sure, though, Küchelbecker's "report" is intended to tell readers what the palaces were like, and we should not be surprised at his impulse to list. At the same time, his list has a specific character. Rather than being just an itemization of unrelated objects, it proceeds under an overarching categorical heading explicit in his first line describing the Belvedere: "Among the buildings west of the city we find that one which easily surpasses all the other palaces, whether they be inside or outside the city. Such is the incomparable palace of the greatest hero of our time, namely, that of Prince Eugene of Savoy."[28]

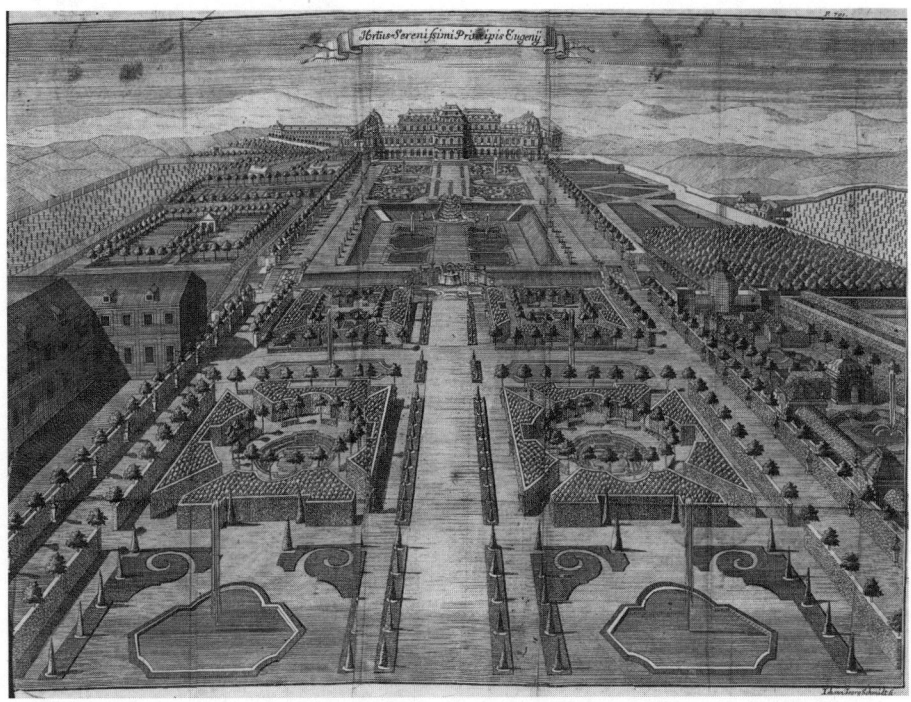

Fig. 8. Johann Küchelbecker, View of the Belvedere gardens. *Allerneueste Nachricht von Romisch-Kayserl. Hofe* (Hanover, 1730).

The listed objects, then, are the collection *of* Eugene—a hero in a time of gigantic heroes—and this single fact is the resounding redundancy throughout the whole of the Belvedere complex. It is as explicit in Eugene's fabulously worked initials beneath heraldic lions in the Great Main Gate as it is implicit in the architecture of the Upper Belvedere's two-story-high, oblong-octagonal "Open Room" with its frescoed ceiling, red marble floor, and vista down across the gardens and Lower Belvedere to the city beyond. One of the illustrations from the Küchelbecker edition (fig. 8) shows a bird's-eye view from above the Lower Belvedere looking across the gardens to the upper building. From the "heavier," higher-hedged geometrically patterned beds—knots—in the foreground, our eyes move along the straight topiary-lined paths toward the Upper Belvedere. Passing the lower cascade with its solid appearance, we slowly enter a distinctly more ethereal world. The main cascade seems to be made of glass, while the upper parterres flatten out into tightly controlled arabesques. The movement, as one author put it, progresses from the "elemental" and "earthly" to the "godly upland watched over by sphinxes."[29] Importantly, Girard, arriving after the preliminary work on the gardens had been completed by Hildebrandt, "reversed their layout so that they could be viewed sloping downward rather than upward."[30]

Eugene's formal garden was a collection of patterns, compositions of color and line composed of flowers and low shrubs. But the prince also amassed specific items. Indeed, he owned more than two thousand different species of plants and trees, most of which were in the "Little Paradise Garden" and more particularly in its Orangerie.[31] It is important to note, however, that Eugene's paradise garden was very different from the ideal paradise garden of a century earlier. The latter was a walled or high, hedge-enclosed square divided into four equal parts which emphasized vertical depth—often with a central fountain—equally with the horizontal, thus creating a surrounding autonomous world. Eugene's "paradise," in contrast, echoed his formal garden by stressing the drawn-out rectangle over the square and thus emphasizing the horizontal control of the observer over the vertical independence of the self-contained paradise.[32] The Orangerie, "Little Paradise Garden," and winter gardens in the Lower Belvedere were pleasure gardens for the prince, where he and his fellow plant enthusiasts such as Friedrich Karl von Schonborn could marvel at the beauty and variety of indigenous and foreign plants. Animals, plants, and even books gained attributes beyond themselves, however, once they had been procured for Eugene: they literally became "of Eugene." Upon receipt his books were sent away to be rebound in rich Moroccan leather. The title of each volume was impressed in gold on the spine beside an arabesqued *E*. The prince's shield was similarly embossed on the front and back covers, and the volumes were ordered through a system of color coding: history and fiction were bound in dark red, theology and law in dark blue, and natural science in yellow.[33]

This presence of Eugene, his ownership, is no less apparent in the engraving by Salomon Kleiner of the view of the menagerie from the second-floor balcony of the Upper Belvedere (fig. 9). This balcony and the living room behind it constitute a typical threshold in the Belvedere. The room, decorated with a colorfully inlaid wood floor, mirrors, and four tables of brown-veined black marble, was a controlled human environment. Upon leaving it and stepping out onto the balcony, however, Eugene did not enter into a completely different world—one, for example, of violent nature. On the contrary, between

Fig. 9. Salomon Kleiner, View of the Belvedere menagerie, ca. 1734. From Hans Aurenhammer, *The Belvedere in Vienna: Ten Engravings and a Description by Contemporaries
of Prince Eugene* (Vienna: Österreichische Galerie, 1963).

the world outside and world inside the Belvedere there was an unhesitating
continuum, and it is Eugene himself who provides the continuity. Indeed, in
Kleiner's view of the menagerie from the balcony, it is clear that the center of
the collection—the point from which all views radiate—is the prince himself.[34]
Far from our contemporary zoo enclosures, which attempt to hide the presence of people, Eugene's view-controlling rows of chestnut trees that terminate
in enclosures based on human buildings emphatically proclaim that presence.
For his architects and for Eugene himself, it made perfect sense for a lion to
occupy the center court of the menagerie and for this court to find its architectural and philosophical center in Eugene. The dominating ethos in Eugene's
collection was one that emphasized the prince's control and power over his
collected objects.

What of Prince Eugene's collections, then? Are they meant just to be gazed
upon from the heights of the Upper Belvedere in some careful orchestration of
self-aggrandizement? Are they intended simply to embody human domination

over nature? In fact, arguments such as these are not completely out of line, but then Eugene, though certainly one of the more glorious men of his age, did not engage in his collecting merely to put on a fantastic show for awed onlookers. Consider for example, his collection of books. Although it was quite large, a letter from Jean-Baptiste Rousseau indicates that scarcely a book in Eugene's libraries had not been read or at least leafed through by the prince.[35] This was a collection that, by the end of his life, amounted to some 15,000 volumes and more than 230 manuscripts on a broad range of topics. Encyclopedic as his tastes may have been, Eugene seems to have taken a personal, intensive interest in his collection. He actively pursued his collecting during his military campaigns as well as through acquaintances. Agents procured art, plants, animals, and other rarities in Amsterdam (still a hub for the consumer), in London, and in major cities in Italy and France. In March 1720, for example, the prince wrote to a Sicilian commander thanking him for a shipment of animals from Nepal and suggested that any animals obtained in future voyages would be greatly appreciated. In December of that same year he wrote to a Belgian ship's captain to express his gratitude for a number of pheasants that had been delivered.[36] Küchelbecker recorded that Eugene's rare plants had been brought from Florence, Genoa, and Naples and had come from Peru, India, Malabar, and Turkey.[37] The prince had purchased date trees and coffee bushes, musk, camphor, and "dragon trees," for one of which he was reported to have paid seven thousand gulden in Holland.[38] Among the specimens in the menagerie were numerous Indian, African, and American eagles and vultures; Indian oxen, sparrows, deer, and wolves; European wisent, deer, and moufflon; African oxen and boars; North American bison; a couple of lions; a four-horned springbok (visible in the Kleiner view of the menagerie); cranes, pheasants, five ostriches, a cassowary, and the list continues.

The point is that all Prince Eugene's collections functioned on at least two levels. Clearly they pointed to the remarkable accomplishments of the prince, to his power and prestige. Standing in the "Open Room" of the Upper Belvedere and looking out over his formal gardens to the city of Vienna in the distance, or standing on his balcony and looking out over his menagerie, Eugene was master of all he surveyed, and that mastery was written into every aspect of his property. But the prince did not always stay on his balcony; he and his guests came down and looked at his animals more closely, read his books, and spent hours studying the plants in his collection. Eugene, like so many other princely collectors, including the Holy Roman Emperors Charles V and Rudolf II, collected paintings, collected plants, collected thinkers, collected knowledge, and collected animals. These princes also collected power, but the menageries, botanical gardens, cabinets of wonder, picture galleries, and fish ponds these men constructed represent more than "simple" expressions of that power.

The Zoological Gardens of London

If it is important to complicate the understanding of the princely menagerie as more than just another representation of all that is wrong with a decadent and unenlightened aristocracy—the party line of zoo historians now for more than a century—we need, as well, to complicate the notion of the nineteenth-century public zoological garden as some altogether new and, indeed, better form of animal exhibition. The bourgeois zoos of the nineteenth and twentieth centuries, those dreams of European urban elites around the world, were both different from and, importantly, very similar to their princely predecessors. What is at stake around those similarities and differences are, of course, some fundamental definitions of the bourgeois state, but we need not go far into an exploration of bourgeois sensibilities and mentalities to get a sense of the essential qualities of these public zoos. In fact, a very brief examination of the Zoological Gardens of London, which, along with those in Paris, Berlin, Amsterdam, Antwerp, New York, and Philadelphia, were one of the most important models of the period, will clarify the essential features of this "new" kind of animal exhibit.

In her "Memoir of Sir Thomas Stamford Raffles, F.R.S.," Raffles' widow recalled that around 1817 "he meditated the establishment of a society on the principle of the Jardin des Plantes, which finally, on his last return from the East, he succeeded in forming, in 1826, under the title of the Zoological Society of London."[39] Raffles' primary aim was to create a forum in which those interested in specifically zoological topics could study and present scientific papers. Raffles and others insisted that the new society was needed because the Linnaean Society, established in 1777 to cultivate the general study of natural history, had focused too narrowly on botanical studies. In response, the broad objective of Raffles' organization was to advance zoological science in its aspects of classification and description but also to domesticate new animals to the uses of agriculture and recreation. The society's initial goals, therefore, were the "formation of a collection of living animals; a museum of preserved animals, with a collection of comparative anatomy; and a library connected with the subject."[40]

With the founding of the gardens in 1828 in Regent's Park, however, the collection of animals swiftly assumed an added character that was anything but scientific. Raffles' early proposals included the idea that the zoological collection should also both "interest and amuse the public."[41] In the end, this latter quality would become perhaps the most important mandate behind the development of the gardens throughout the century. Indeed, by the mid-nineteenth century, the gardens resembled much more a public place of entertainment than some sort of scientific station. Although originally only fellows of the society and their guests were admitted to the gardens, within a dozen years the general public was admitted on Mondays and Tuesdays for the payment of a shilling each and on other days with payment and a written voucher from a fellow. By the end of the 1840s—even though Sundays and, beginning in 1844, "Promenade" days continued to be set aside for the fellows—the public was admitted Monday through Saturday, paying sixpence on Monday and a shilling the rest of the week; children paid sixpence all week. By 1850, however, the last barriers began to crumble, and the Promenades, with their military bands, exotic entertainments, beautiful clothes, and garden teas, had also been taken over by the general public and had become part of the regular Saturday fanfare at the zoo; they had, indeed, become part of "general admission." In the end, by the second half of the century, the gardens had become a well-established and highly acceptable venue of outdoor public entertainment—something, it seems, between an urban nature park and an amusement park—and this development seems to have been typical of the other major zoological gardens of Europe and the United States.

The basic qualities of this new space are apparent in a series of midcentury lithograph drawings of the gardens executed by Thomas Hosmer Shepherd. His *The Zoological Gardens, Regent's Park* (fig. 10) captures the general feeling of many similar illustrations from the period. At the right rear of this view of the Music Lawn can be seen the Camel House Clock Tower, designed by the zoo's first architect, Decimus Burton, and built in 1828. Down the left of the scene runs the Carnivora Terrace of 1843, with its Ionic pilasters and flat surfaces sur-

Fig. 10. Le Pettit (after Thomas Hosmer Shepherd), *The Zoological Gardens, Regent's Park*. From *Views of Mighty London: Its Environs and Royal Palaces* (London, 1854). Yale Center for British Art, Paul Mellon Collection.

rounding picture-frame cages housing lions, tigers, hyenas, and bears. At the rear of the scene stands the oblong Polar Bear Cage, constructed around 1832, with its high inward-arching steel bars. But more than the animals at this zoological garden, indeed more than the remarkable buildings, the central focus of the lithograph remains the lawn and its human inhabitants. Indeed, the presence of people seems absolutely essential to this work. Friends, couples, and families with well-behaved children walk and talk, stand before the cages studying animals, gather to socialize, and present edibles to an apparently free-roaming pair of elephants. A picture like this was a simple impossibility a century earlier, in the times of Eugene's Belvedere menagerie. Although there were both portraits of Eugene and works depicting views of the Belvedere Palace, in the early decades of the eighteenth century there was neither an anonymous "public" milling around the gardens and menagerie of the Belvedere which could serve as the focus of illustrations nor a tradition of portraiture which would allow the depiction of the prince in the midst of his collections. Thus, the people present in illustrations of the grounds of the Belvedere in the times of Eugene seem to exist only to provide scale to the real focus—the palace itself.

In contrast, the Shepherd illustration of the Music Lawn encapsulates the critical essence of the popular nineteenth-century zoo: this was a place designed by the bourgeoisie for its own education and amusement. The atmosphere of this remarkable new public institution thus encouraged social events such as band concerts and promenades and the presentation of animals in contexts saturated with human references. In a passage that seems in many ways characteristic of the period, for example, a guidebook for the gardens from the early 1860s describes the path from the main entrance to the Carnivora Terrace:

> From the rustic lodges at the north, or main entrance, runs a broad terrace walk, in a straight line onwards, bordered by flowers, shrubs, and trees on each side, and continued at the same level for some distance, over the lower ground, by a handsome viaduct [the Carnivora Terrace], which covers a long range of roomy cages beneath, and in itself forms one of the most striking objects in the Gardens. On this platform, which is balustraded at the sides, the visitor may pause for a moment, to contemplate the extensive view presented of Regent's Park, and the mighty Metropolis beyond. Save its smoke, however, and the mist, or dense air, perpetually hanging over it but little of the latter is visible. Still it is not less present to the imagination's eye, and the contrast is the stronger when compared with the tranquil scene around.[42]

The passage emphasizes the gardens as a place of quiet repose in the heart of an industrial city. Underscoring the smoke and thick air seemingly "hanging" over the city, the passage suggests that this air is somehow magically lifted at the gardens, a place where thoughtful people could find an opportunity to contemplate the striking contrasts between dense urban sprawl and hushed nature. Of course, the animals could not always be relied upon to cooperate with this idyll. The raucous sounds of the bird houses—made all the worse through their frequent construction as glassed-in conservatories—the smells of the great cats, the inopportune matings, and the sometimes pathetic conditions of the captive animals all drew regular criticism. In a typical letter to a director of one of the bourgeois zoos (in this case William Hornaday of the New York Zoological Society's Bronx Zoo), John P. Haines, president of the New York Chapter of the American Society for the Prevention of Cruelty to Animals, wrote, for instance, with a common lack of irony about the importance that the collections not be in any way depressing for the *human* visitor:

Dear Sir

A friend of animals called at this office yesterday, and after highly complimenting the manner in which the animals are housed at the NYZS, said that she had been distressed by one thing she saw, and to which she asked us to call your attention. She said that she noticed in the bird house, in the cage devoted to birds indigenous to this section, a poor robin which was apparently in a droopy and sickly condition. It seemed to our complainant that in view of the prevalence of the common robin, and the ease with which a specimen can be obtained, the society might at least confine a healthy bird, if it is considered necessary to keep a robin in confinement.

We refer this complaint to you, knowing that you will do what is proper in the premises.

Yours very truly,

JPH

President[43]

Characteristically, the focus of this complaint is not that it might be a good idea to see if someone could do something for the sick robin but that it is somehow wrong to exhibit a sick robin in a cage. In the end, this reasoning stems from the zoological gardens being imagined as places of amusement; there is, apparently, little fun to be had in "drooping" animals. Of course, "do not feed the animals" signs were as yet unimagined in late-nineteenth- and early-twentieth-century gardens. Indeed, feeding the animals, either by oneself or through a zookeeper, seems to have been a central part of visits to the zoo, and the extended arm holding out food to the animals is perhaps the quintessential gesture of these places. This is as clear in the Shepherd illustration as it is in innumerable other illustrations depicting gardens in the period. It is equally clear in the idea of the bear pit as well as in the ritualized feedings of the great cats, the seals, the penguins, and other animals that became a focus of activities at gardens.[44]

Throughout the bourgeois zoo we see the presence of "the public." A way of imagining animals in human contexts pervaded the Berlin Zoological Gardens' magnificent buildings—such as the huge Elephant Pagoda, built in supposed imitation of a Hindu temple and boasting yellow, brown, and blue domes (fig. 11), or the Ostrich House, designed to resemble an Egyptian temple with paintings of "ancient" figures of men and birds covering both the interior and exterior walls—as persistently as it did at London's neoclassical Carnivora Terrace. Indeed, the ornate conservatories, with their apparently

Fig. 11. The Elephant Pagoda at the Berlin Zoo. From *Der Berliner Zoo im Spiegel seiner Bauten, 1841–1989,* ed. Heinz-Georg Klös and Ursula Klös (Berlin: Heenemann, 1990). Courtesy, Zoo Berlin.

delicately wrought cages, which stood in hot, humid light and were assailed from all sides by a talking, poking, and feeding public, owed their design almost entirely to cultural expectations that exalted the presence of civilized humans in a world of beasts. The avowed simple desire to keep a scientific collection for the education of the public and the research of scientists simply was not part of the equation in designing these places. The London gardens' near obsession with parrots, for example, and the public delight in the Parrot Walk—a special attraction on pleasant days in which birds were tethered to rows of hanging stands down which the visitors could stroll—had little to do with nineteenth- or twentieth-century scientific interest in the birds. Thus, the argument that the almost fifteen hundred different species and subspecies of birds represented in the London gardens' collection between 1831 and 1900 were obtained almost solely for scientific reasons—for use in studies comparing the differences and similarities between the physical structures of different specimens—misses the mark.[45] The other side of the Zoological Society members' studies of comparative morphology was the aesthetic pleasure of walking down a row of parrots; the animals appealed to the visual, tactile, and aural desires of the gardens' patrons.[46]

Menageries and Zoos

Fundamentally, the nineteenth-century European and American zoological gardens, with their garden teas and concerts alongside science and education, were places designed to celebrate the tasks of enlightened and bourgeois progress in the world. They showcased the optimism, power, and ambitions of the new bourgeois elite, just as the princely menageries showcased the optimism, power, and ambitions of an older aristocracy. With this said, it should still be clear to anyone who compares Salomon Kleiner's view of the menagerie at Prince Eugene's Belvedere palace (fig. 9) with an illustration of feeding the hippo at the London Zoo (fig. 12) that the two places are very different. When one looks at these pictures, it seems easy enough to agree with the zoo historians that the closing of the princely menageries and the opening of the public

Fig. 12. Margaret Tarrant, Feeding the hippopotamus at the London Zoo. From Harry Golding, *Zoo Days* (London: Ward, Lock & Co., 1919).

zoological gardens represented a fundamental reorientation in the exhibition of animals. But what is the change?

The claim of the institutional zoo historians has been that the public zoological gardens of the nineteenth and twentieth centuries fundamentally differ from earlier collections of animals because they emphasize science, education, recreation, and conservation and are no longer places of simple curiosity or the expression of some sort of cultural or personal power. It is, of course, true that the owners of the princely collections usually had no interest whatsoever in public education or conservation. With that said, it is also true that the history of education and conservation in zoological gardens over the past hundred and fifty years has been mixed at best. While it is clear, for example, that the zoo seems to present a remarkable opportunity to educate large sectors of the public about the lives of animals, we all know that the gardens have always struggled in their efforts to educate. However much zoos employ educational signage and special exhibits, most people still go to the zoo because it is fun, not because it is educational. Indeed, when the British music-hall artist "the Great Vance" coined the word *zoo* in his 1869 song "Walking in the Zoo," the word clearly connoted amusement and pleasure, not education. Similarly, although it is true that the coordinated efforts of zoological gardens over the past century have resulted in the preservation of a couple of dozen animal species from extinction, it is also true that there are quite a few cases in which the efforts of nineteenth- and twentieth-century zoological gardens to obtain and exhibit—or protect—certain particularly endangered animals have contributed to difficulties faced by that species.[47]

As for the other claimed unique attributes of the zoological gardens—recreation and science—there is every reason to believe that these were vital elements of the earlier menageries. To be sure, recreation at a princely collection was afforded to a distinct and very limited population, but this has also been the case for the premier nineteenth- and twentieth-century zoos, which have always been the domain of a limited public, even if that public has changed somewhat over the past century. At the same time, substantial scientific work found a place in many animal collections before the Jardin des Plantes and the Zoological Gardens of London. Science was deeply rooted in such collections as those of Eugene of Savoy, the Holy Roman Emperor Rudolf II, and William of Orange. In addition, people interested in science and the lives of animals regularly attended the visits of traveling animal exhibits. As we have seen, we can thank a traveling exhibit in the mid-eighteenth century for finally correcting Western ideas about the appearance of the Indian rhinoceros, and almost our entire understanding of the appearance of the dodo can be credited to the Flemish/Dutch artist Roelandt Savery, who, it seems, had the opportunity to study the bird at the princely collection at The Hague in 1626.

In response to the final claim that animals in menageries were kept only to aggrandize their owners, it is clear from the example of Eugene that menagerie animals certainly could become markers for the accomplishments of their owners. At the same time, however, it is also clear that modern Western zoological gardens, from the nineteenth century to today, often act out quite explicitly the political, imperial, or educational claims of the current elites—they are the grounds, for example, in which we attempt to teach appropriate judgments about the exotic world. This is as obvious in the early-twentieth-century exhibition of the pygmy Ota Benga at the Bronx Zoo as it is in the current didacticism of zoos about the necessity to stop the destruction of the rain forests by local farmers and ranchers.[48] Indeed, that so many conservation programs at zoological gardens—including such things as "rain forest meters," whereby visitors can put a nickel into an adapted parking meter to save a small piece of rain forest somewhere—are based on the idea of purchasing land in faraway places for protection suggests that the imperial impulses of the great nineteenth-century zoos are not as distant from contemporary zoos as we may want to believe. In short, the claims that zoo historians have traditionally made about the progressive changes wrought by zoological gardens should be accepted only with great caution.

It is an old game for historians to point out that commonly held distinctions between different historical periods do not usually make a great deal of sense. If talking about a progression from menageries to zoological gardens to conservation centers is not terribly meaningful, however, then how are we to understand the changes we can, in fact, see over the course of the centuries in the ways animals have been presented? To begin an answer, let's consider the 1864 painting by Paul Meyerheim entitled *Menagerie*, which is now in the remarkable collection of zoo-historical materials owned by Werner Kourist and which details a fairly unremarkable sight in the history of traveling menageries (fig. 13). Technically and stylistically there is nothing terribly important about the work; nevertheless, its content remains thoroughly intriguing. At the right rear of the scene we can make out a portion of a railway wagon used to transport the

Fig. 13. Paul Meyerheim, *Menagerie*, 1864. Courtesy of Werner Kourist, Linz am Rhein.

animals from town to town. Along a side of the tent the cages are lined up edge to edge with smaller crates stacked one above the other. A few skins of dead animals are draped casually over the side of a lion cage out of which hangs some loose hay and the indifferent attention of a heavily maned lion. Above the group of onlookers two South American macaws contest intrusions into each other's territories, while a few feet away an African gray parrot sits quietly alone on his hanging stand. At the left front of the picture we see a Bactrian camel beginning to molt its heavy winter coat. On top of a crate stands the impresario, his body wrapped in the coil of a python. At his feet is a small tin container holding warm water that, along with the wool blanket, is intended to protect the animal from the cold weather. In the foreground a pelican snaps wildly, perhaps having been frightened by some too inquisitive children. The crowd of onlookers is a mixed group both in age and in profession; a gentleman in a top hat at the rear of the scene seems the least interested in the immediate action and has turned his back to the painter while he looks at a distance into the smaller cages. Overlooking the arena an organ grinder's monkey wearing a cap is tied with a rope attached to a harness around its waist to one of the posts holding up the tent. The monkey, perhaps frightened by the appearance of the python, has jumped from his little platform to the supporting beam from which hang the bird stands. On the beam, and barely legible, we can read, "Quäle ni ein Thier zum Scherz, denn es fühlt" (Never torture an animal in fun, because it feels).

In a way, Meyerheim's painting is about confrontation: on one side, the exotic world of lions and snakes, the extravagantly costumed and loud voiced; on the other, the world of civilization and Europe, of rapt awe and studied indifference, of red umbrellas and green locomotives. Somehow the pelican, sort of an advanced guard of the exotic, has crossed the lines, has left the refuge of his fellows, and now, on the defensive, claps his huge beak and tries awkwardly to return to his old comrades. The curious, quiet, and tentative visitors to the tent appear somehow to have stumbled by accident into a bazaar, where they are greeted by strange sounds, smells, and sights. A small boy in the foreground, hardly noticing the animals, is dazzled by the huge, powerful man before him. Standing on a crate, the orator is a spectacle: his massive legs with boots rising to his knees are spread apart commandingly, and with his red vest, dark skin, and powerful arms, he presents a stark contrast to the small, drab, and seemingly feeble men surrounding the boy. While the boy stares at the explicator, an older, bearded man stares directly at the snake, which itself seems to be intrigued by the audience. The eyes of the woman between the boy and the man seem to rest on the snake's heavy coil around the impresario's chest; could this snake be the killer of Laocoön, the orator seems to ask? The audience's response appears to be a mix of fear and wonder. This small group of animals that had been brought to Europe from the corners of the globe points in its small way to a very large story about a fascination in Germany, and indeed in Europe and North America more generally, with distant frontiers—frontiers of the state, but also the frontiers of knowledge and civilization. One has to realize when looking at a painting like this that there are many different kinds of stories to tell about it. The relatively uninteresting ones have to do with positioning the work's content within a progressivist history of becoming enlightened about exotic animals. The more interesting ones, at least to my mind, focus on reading the many layers of narrative in the picture, seeing it as a historical event full of detail.

Reading layers of narrative seems like a good approach in trying to figure out the meaning and significance of both historical and contemporary zoological collections, and perhaps of Carl Hagenbeck's exhibitions in particular. Visiting Hagenbeck's Animal Park in its first days, for example, Friedrich Katt, a correspondent reporting to the journal of the association of German zoological gardens, noted that Hagenbeck had "occupied himself with completely different issues from those of the scientifically oriented zoologist who [stood] at the head of the older zoological gardens." Hagenbeck's past as an animal dealer and trainer, Katt argued, had led to his creating "something at essence popular, an animal show for the visiting public and the animal buyer, something therefore, totally different from a zoological garden as that concept [had been] generally understood."[49] While admitting that the blatant "theatricality" of the project seemed to "deviate" from the traditional gardens, Katt conceded, "One has nevertheless seen something unusual, something gigantic, when one leaves," and further, "Hagenbeck's enterprise has assured itself a place in the history of keeping animals as an entirely new kind of zoological institute."[50] Hagenbeck's park was something different, more exaggerated and more exciting than any "normal" zoological garden. Recognizing essentially two distinctive elements of the park—commerce and theater—Katt was overcome with the impressive display and somewhat baffled by it at the same time.

Indeed, the way in which Hagenbeck's exhibits created for the visitors a theatrical illusion of freedom for the animals was noted, sometimes admired, often dismissed, and inevitably emulated by zoological gardens all over the world ever since. The point of this kind of illusion—the point of this kind of theater—is, however, narrative. These exhibits are meant to tell a story. As we shall see, though, that story is often difficult to pin down. On its most basic level, of course, the illusion created a sort of idealized world in which the structure of the zoo itself disappeared. This idea is suggested by Ludwig Zukowsky, one of Hagenbeck's assistants. According to Zukowsky, Hagenbeck had wanted "to create an animal paradise which would show animals from all lands and climatological zones in a manner suitable to their life conditions, not from behind bars and fences, but in apparent total freedom. This paradise would also exhibit people of all colors. It would be a nature sanctuary in the most truthful sense, a world in miniature; and thousands of visitors would be able to make a danger-free trip around the world and stroll peacefully under palms."[51] Not simply an enterprise rooted in education, conservation, science, and the improvement of the urban landscape, or even an expression of imperial ambitions, Hagenbeck's park was a fantasy world that was openly theatrical, and the dramas that played there were often monumental.

Unusual as Hagenbeck's park was and to a lesser degree still is, however, it points to the remarkable persistence—both long before and after the park's creation—of what became the Hagenbeck motif. From Roman arenas to eighteenth-century noble collections, to nineteenth-century temples to nature, and to our contemporary Sea Worlds and Animal Kingdoms, the element of dramatic entertainment reflected so well in Hagenbeck's park has survived remarkably unchanged through the centuries of exhibitions of exotic animals. To be clear, the Zoological Gardens of Berlin and its predecessors were not simply expressions of dramatic or perhaps moral theater. Rather, the continuous element in zoological collections—including that of Eugene of Savoy and the Zoological Society of London—has been the simple and yet powerful idea of display. Indeed, in order to understand the full importance of such cultural productions, we must begin to interpret the individual and distinctive stories that the exhibition of particular enclosures, species, or even particular animals can tell us. Hagenbeck's animal trade, Hagenbeck's exhibitions of people, Hagenbeck's circus, and Hagenbeck's Animal Park are not simply expressions of the colonial power of Germany, the commercial power of Hamburg, the imperial power of anthropology, ethnology, or biology, the aesthetic solutions to a central ideological debate about captivity, or any other overarching analytical category. "Hagenbeck," as an acquaintance of mine pointed out to me as he recalled long-cherished stories of his grandmother, who had been kissed on the cheek by one of Hagenbeck's "Africans," is a sort of mythic figure, and as such, "Hagenbeck's" is about telling stories, partly true, partly fictional, which changed with seasons, with political and economic climates, with source material, and with directorship. It is to these stories that we must now turn.

One of the most remarkable stories about the firm of Carl Hagenbeck—one to which we will return throughout the rest of this book—is Franz Kafka's 1917 "A Report to an Academy." In this short story, Kafka explores themes of captivity, freedom, and art through the voice of a "civilized ape." The story is written as an address by the ape, Red Peter, to the members of a scientific society. The academy had requested an account of the life he had formerly led as an ape, but Red Peter insists that such a project remains impossible because, although his existence among—and as one of—the civilized has consisted of only a few short years, he has nevertheless advanced so far from his former life that his memories of that earlier time have all but disappeared. Red Peter concludes, "Your life as apes, gentlemen, insofar as something of that kind lies behind you, cannot be farther removed from you than mine is from me."[1]

But if Red Peter is unable to recall substantially his life as an ape, he is nevertheless prepared to relate how he left his apeness behind to become human. The steps to that accomplishment began with his capture by a "hunting expedition sent out by the firm of Hagenbeck." Red Peter's actual memories begin, however, "between the decks in the Hagenbeck steamer inside a . . . three-sided cage nailed to a locker." "The whole construction was too low for me to stand up in," he complains, "and too narrow to sit down in. So I had to squat with my knees bent and trembling all the time, and also, since probably for a time I wished to see no one, and stay in the dark, my face was turned toward the locker while the bars of the cage cut into my flesh behind."[2] Red Peter soon concluded that by becoming human he could escape his cage, even if he could never regain his freedom. The steps to his humanity were fairly straightforward: learn to shake hands, learn to spit, learn to drink, learn to smoke, learn to speak, learn, in short, to play a human on the variety stage. On a simple level, Red Peter, the artist, finds his "way out," as he calls it, through conforming to and exceeding expectations.

While admittedly it is the general theme of the relation between art and freedom which drives the narrative of this story—a theme taken up again by Kafka in such other animal works from the early 1920s as "Josephine the Singer," "The Burrow," and "Investigations of a Dog"—it is important to remember that in "A Report" Kafka is also, in fact, discussing the capture, transport, and exhibition of exotic animals. This would be a point of only passing interest were it not for Kafka's repeated references in the short story to the name "Hagenbeck"—a name that, perhaps more than any other in the period in Germany, could bring forth images of the wild, the exotic, and the animal. As the dramatist Carl Zuckmayer put it at the end of the 1940s, "When a child of the South of Germany thinks of Hamburg, he paints a picture in his mind of a small town of red-brick buildings directly on the open sea, and encircled on all other sides by a huge and magic kingdom, Hagenbeck. Hagenbeck is not a proper name, but rather, like Alaska or the Wild West, the expression of a mysterious, unexplored land, where one yearns for adventure."[3] In this chapter we turn to the early history of the firm of Carl Hagenbeck and examine the development of one of the central elements of the "idea" of Hagenbeck—the idea that grounds Kafka's essay: the experiences of animal catchers and animals caught in far-off places of the world.

Beginning with Seals

The beginning of the Hagenbeck exotic animal business can be traced to March 1848.[4] One day early in the month, contracted sturgeon fishermen employed by Carl Hagenbeck's father, the fishmonger Claus Gottfried Carl Hagenbeck (1810–87), had been working the lower Elbe and returned with six seals that had become caught in their nets.[5] Carl Hagenbeck recorded in his memoirs, *Von Tieren und Menschen* (Of beasts and men), that his father "came to the lucky idea of letting people see the animals for the price of an entry fee, and to this end put them in two large wooden basins on the Spielbudenplatz in St. Pauli." According to Carl, the success of the family's first attempt at exhibiting nondomestic animals went far beyond expectations. After the Hamburg public's interest in the seals began to wane, Carl Hagenbeck Sr. was induced through business acquaintances in Berlin to take the seals to the Prussian capital to be exhibited again, and in the middle of the violent domestic political upheavals of the Revolution of 1848, he made a neat profit showing his seals to anyone interested before selling them to another entrepreneur. As the younger Hagenbeck noted, "That was the beginning of our animal business."[6]

After the success with the seals, Carl Hagenbeck Sr. increased his forays into the exotic animal market. In addition to regularly exhibiting and selling seals, he collected all kinds of animals for his house menagerie. Hyenas, opossums, and a variety of birds and monkeys found temporary quarters in the house; the father even bought a raccoon, which escaped on its way from Bremerhaven to Hamburg and evidently lived happily for a year and a half marauding local farms before ending up as someone's extraordinary hunting story. By the early 1850s, Carl Hagenbeck Sr. regularly bought animals that arrived in ships in the port of Hamburg. Although there were already a number of exotic animal dealers in Hamburg who traded in birds, monkeys, and other small mammals, unusual reptiles, and such popular collectors' items as butterflies and beetles, Carl Hagenbeck Sr. was often willing to buy what others rejected—especially the larger animals for which the market was relatively weak. Other than the zoological gardens in Berlin and Frankfurt—both still in their infancy—and the traveling circuses and menageries, there was little call in 1850s Germany for such animals as zebras, antelopes, ostriches, bears, and the great cats. Nevertheless, Carl Hagenbeck Sr. had a taste for the grand, and he soon gained a reputation as a key animal dealer.

Despite some sensational purchases and sales, however, throughout the 1850s the Hagenbeck exotic animal business was more a hobby to Carl Hagenbeck Sr. than a business. His primary enterprise remained buying, selling, and processing fish, and with the increasing popularity of sturgeon caviar, that business was doing well. In contrast, the riskier animal business, according to his son, presented recurring losses. Quite simply, through inexperienced handling many animals died before they could be sold or profitably displayed. Despite the risk, young Carl—having completed his formal education at fifteen years of age—had decided to devote himself entirely to the animal business.[7] His younger siblings followed in his steps: his sisters Luise and Christiane took over the care of the birds, for example, and his brother Wilhelm helped with the other animals and experimented with new training methods. The deals that the younger Carl began to close were becoming larger, and soon he began to cut away at the limited markets of his competitors, including those of his main rival, the English firm of Charles and William Jamrach, which, along with a number of smaller English companies, had controlled the greatest share of the market for large exotic animals in Europe.[8] Attending the annual fall animal auction at the gardens in Antwerp in 1862, for example, the eighteen-year-old Hagenbeck, using his contacts with the directors of the zoological gardens in Cologne and Dresden, quickly sealed contracts with the directors of the zoological gardens in Paris, Rotterdam, and Amsterdam. Shortly thereafter, the young Hagenbeck purchased the traveling menagerie of Christian Renz, including an evidently magnificent North African lion, an Arctic wolf, a jaguar, and several panthers. He then sold most of the animals immediately on the spot at the fair in Krefeld, bringing the family business a profit of some two thousand talers.[9]

In 1863, Carl's father responded to the growth of the family animal business by moving to a larger property at Spielbudenplatz No. 19—a residence that had previously housed the menagerie of Jacob Gerhard Gotthold Jamrach (the father of Charles Jamrach), who had moved the center of his family's animal business to London and Liverpool.[10] The house and its rear court were illustrated and described by Heinrich Leutemann in 1866 for the bourgeois magazine *Daheim* (fig. 14).[11] On the street side of the house were two shops, one rented by a shoemaker and the other used by the Hagenbecks as a bird shop, a branch of the business which was taken over completely by Carl's sister Christiane in 1870. As with all the properties on Spielbudenplatz, behind the house was a large court where most of the menagerie was housed. Leutemann's illustration presents an assortment of presumably fairly common sights at the Hagenbecks: a sea captain has just brought a small animal, and Carl unpacks it from its crate. Near the front rests the perennial wooden basin used for displaying the seals. Two men to the left struggle with an apparently stubborn reindeer as a polar bear looks out from its cage located in the right rear. A monkey rides on the back of one of the unpackers, a peccary with free rein stands in the foreground, and a dog and a peafowl look on while other assorted larger birds carry on in cages in the rear. At the front of the scene more birds look for the food thrown by one of Carl's sisters, who is seen holding a small basket. The image, a composite of the many activities taking place at the Hagenbeck *Thierhandel* (animal business), presents well what one might have expected to see upon visiting the small business and has become one of the central icons of the family history.

By the mid-1860s, Hagenbeck's business began slowly to change its char-

Fig. 14. The Hagenbeck animal business in 1866. Courtesy of Hagenbeck's Tierpark.

acter. On the one hand, the company continued to purchase the haphazardly arriving animals brought by the ships docking at the Hamburg port. In 1863, for example, a quartermaster of the HAPAG line brought the five large bears that had belonged to Grizzly Adams—the onetime trapper and later impresario—to Hamburg after having purchased them in the Adams estate sale.[12] The two grizzly bears, one black bear, and two brown bears were too much for any dealer in Hamburg to manage other than the Hagenbecks, so Carl and his father purchased the animals for an unusually low price. But while incoming ships continued to be met by the Hagenbecks and their representatives, Carl also attempted to secure more regular and dependable lines of supply by building contacts in England, France, Holland, and Belgium. Still, at this point, most of the business consisted of buying and reselling animals that had already been brought to Europe and were being sold by menageries, circuses, zoological gardens, or other dealers. Recalling that in 1864 he made his first trip to England to purchase a South American giant anteater, Carl noted that his "dependence on the London animal market ceased only later after the founding of the German Empire and the growth in German overseas relations."[13]

Although it is clear that the unification of Germany and the drive toward and eventual acquisition of formal colonies did give a boost to the Hagenbeck animal business, it is also clear that by 1864—twenty years before Germany really became a player in the race for colonies—Carl Hagenbeck had already made his first step toward the elimination of middlemen and the acquisition of his own sources of supply from what would eventually amount to outposts all over the world. Two years before, in 1862, an Italian named Lorenzo Casanova, who had found his way to Africa after losing his dog-and-monkey show in St. Petersburg to fire in 1859, had brought to Europe a large collection of animals which he had obtained from the Egyptian Sudan. Hagenbeck recalled that the transport had "consisted of six giraffes, an elephant, and many other rare animals." Casanova had encountered some difficulty getting rid of the animals, however, and had sold them cheaply to a traveling menagerie. Receiving a telegram in the summer of 1864 that Casanova was returning with more animals, Carl immediately set out to meet the man, purchase whatever was available, and, most important, sign an exclusive contract with the catcher. The 1864 shipment was small, consisting only of "two young lions, three striped hyenas, a collection of very beautiful large monkeys, [and] a number of birds," but Hagenbeck bought the lot, and Casanova signed the contract.[14]

A year later, in July 1865, Casanova returned with the first contracted shipment of animals, consisting of "three beautiful African elephants, several young lions, a number of hyenas and leopards, young antelopes, gazelles, and ostriches."[15] Similar contracts were extended during the following years to other adventurers, and by the late 1860s, Carl Hagenbeck increasingly seemed to have both the best supply of animals and the most respectable clientele of buyers. By the 1870s, with the completion of the Suez Canal and the growing prominence of German trade and Hamburg businesses in Africa and the Far East, Carl Hagenbeck's business was rapidly developing into the leading firm in the international trade in exotic animals. With almost every year the shipments of animals grew in frequency and size. In the spring of 1870, for example, Hagenbeck received telegrams from Africa indicating that two of his agents there, Casanova and Migoletti, were on their way to Suez and Alexandria, respectively. Casanova's report included the request that Hagenbeck meet him in Suez because he had fallen fatally ill. Carl left for Suez with his eighteen-year-old brother, Dietrich, who himself died three years later in Zanzibar during a catching expedition. Arriving at the train station in Suez, Hagenbeck found a number of animals already loaded onto wagons, but it was the scene at Casanova's hotel which really caught his attention. Carl noted in his memoirs, "I will never forget the unique scene with which I was greeted when we entered the court of the hotel. Had a painter seen this sight, he would perhaps have immortalized it under the title, 'chained wilderness.' Elephants and giraffes, antelopes and buffalo were bound to the palm trees. In the background sixteen full-grown ostriches ran about freely, and in sixty crates moved lions, leopards, cheetahs, thirty spotted hyenas, jackals, lynx, serval cats, monkeys, marabou storks, rhinoceroses, birds, and a large number of birds of prey."[16] With some difficulty, Hagenbeck and his brother took over Casanova's caravan and managed to ship the animals by train to Alexandria, where they were joined by the other transport. The entire shipment of animals, which boarded a ship bound for Trieste, represented, according to Carl, "the largest animal transport that to that point had been brought to Europe."[17] Among the animals were "a rhinoceros, five elephants, two warthogs, four aardvarks, fourteen giraffes, twelve antelopes and gazelles, four wild Nubian buffalo, sixty large and small beasts of prey, including thirty spotted and striped hyenas, seven young lions, eight leopards and cheetahs as well as several other wild cats, etc. Beyond all of those, came twenty-six African ostriches, among which were sixteen full-grown birds. . . . The transport was completed by twenty large crates with monkeys and birds, as well as seventy-two Nubian milk-goats, a wandering dairy which provided milk for our young animals."[18] This impressive shipment more or less sealed Hagenbeck's reputation as the leading figure in the international trade in exotic animals. En route to Hamburg, Carl dropped off animals in Vienna, Dresden, and Berlin, the delivery at the last being the largest during the first year of Heinrich Bodinus's tenure as director of the zoo—a time during which the number of species at the zoo almost doubled and the number of specimens tripled.[19]

The Catchers

If many of the elements of Hagenbeck's reputation were more or less in place by the early 1870s, the image of animal catching which rested behind Kafka's story of Red Peter still lay decades ahead. It would be some time before Hagenbeck's name could be so easily connected to the middle passage of an ape bound for Europe on a large steamer, but to begin to get a sense for the background of that image, we must turn away from Hagenbeck for a moment and direct our attention to the animal catchers themselves and how those hunters presented animal catching to the European and American public.

First of all, it is important to recognize that there were many different kinds of Hagenbeck animal catchers. Among the participants were professional commercial hunters, such as the elephant hunter Hans Hermann Schomburgk and the walrus and whale hunter Ole Hansen, who took up sideline interests in catching the young of the animals they hunted.[20] There were naturalists and explorers, such as the photographer Carl Schillings, who occasionally found themselves collecting live specimens.[21] Also included were soldiers of the colonial armies, such as Hans Dominik, and settlers in German and other colonial outposts, such as Carl Hagenbeck's half brother John, who found themselves well situated to catch or purchase animals for Hagenbeck and other dealers.[22] There was also a group of more or less independent entrepreneurs, such as Casanova, Bernard Cohn, and Josef Menges, who found their peculiar niche in the animal trade fairly early on, and often by accident, but who were particularly adept at working with indigenous peoples to obtain animals and often had long-term relationships with Hagenbeck.[23] Finally, there were the professional catchers, including Christoph and Elisabeth Schulz, Jürgen Johannsen, and Wihelm Grieger, whom Hagenbeck himself, at least initially, outfitted.[24] The backgrounds of these people were, of course, also quite varied. Hans Schomburgk, for example, was the nephew of Sir Robert Schomburgk, the explorer of Guayana and later English consul general in Siam. A man of education and some, if limited, means, Schomburgk took up commercial elephant hunting to finance his interests in natural history.[25] After success in catching a young elephant that he later sold to Hagenbeck, however, Schomburgk gave up elephant hunting to become an animal catcher and later a photographer and filmmaker. Josef Menges, on the other hand, though clearly a first-rate explorer and adventurer, had little interest in hunting and little time for studies in natural history. After arriving in Africa in the 1870s and participating in Charles Gordon's explorations of the White Nile, Menges stayed on in Africa and worked for Hagenbeck from 1876 to 1890 in the Sudan, Somalia, and Sri Lanka before struggling on more or less independently, apparently often at the point of bankruptcy, through to around 1909.[26] In short, all sorts of people ended up catching animals for all sorts of reasons. Beside elite hunter-naturalists like Carl Schillings, a man much admired by Theodore Roosevelt,[27] there were adventurers of lesser means, such as Christoph Schulz, who initially went to Africa because of a desire for adventure but soon found substantial support from Hagenbeck's firm.

During the fifty years before World War I, as different types of animal catchers entered the business, the practice of animal catching in colonial lands evolved dramatically. What had really only begun to take off in the mid-nineteenth century in response to the proliferation of zoological gardens, circuses, and hunting parks was, by the beginning of the war, a rapidly maturing industry. Despite all the different locales and traders, the evolution of the business can be summarized in three basic points. First, the trade had begun with almost complete dependence on both indigenous animal catchers and indigenous middlemen, with European traders purchasing only those animals that reached coastal and later inland trading stations, but by the end of the nineteenth century Europeans had taken over the management, step by step, of each stage of the process. Thus, for example, whereas Hagenbeck's first "travelers," such as Casanova and Menges, bypassed coastal middlemen by setting up their own trading stations in the interior, where they dealt directly with indigenous catchers, later Hagenbeck catchers, such as John Hagenbeck and Christoph Schulz, took over the catching process as well, albeit relying on the substantial labor of local peoples. Second, although throughout this period a number of remarkable catches (for example, the first pygmy hippopotamus, the first German colonial elephant, the first walruses), and indeed full collections, were organized by a varied group of people whose primary interests were not in catching animals—naturalists, colonial administrators, and various kinds of hunters— the catch soon became dominated by people whose primary occupation was collecting animals for zoos and circuses. Thus, although the rhinoceros collected by Carl Schillings in German East Africa in 1903 represented a sensational catch, the large corralled collection of zebras at the Hagenbeck-Schulz farm in German East Africa (fig. 15), or Wilhelm Grieger's heavily financed expeditions beginning in 1901 to capture specimens of the Przewalski horse in Mongolia, came to represent the more significant part of the Hagenbeck business.[28] Third, whereas during most of the nineteenth century portrayals of animal catching in popular articles and books often display an enthusiasm for describing the frequently bloody and destructive methods employed in most catches, by the end of this period accounts began systematically to avoid discussions of the "losses" both during and after the catch. By 1914, a code of catching had begun to develop—partly in response to popular criticism and partly as a result of the professionalization of the practice—which argued both that

Fig. 15. Zebra Corral, German East Africa, ca. 1910. From Ellen Velvin, *From Jungle to Zoo* (New York: Moffat, 1915).

catchers embodied a deep love of nature and animals and that their activities as catchers represented an integral part of larger projects of conservation and education. In short, the animal-catching business was taken over by colonialists and professionalized in much the same way that other colonial industries had been. These general developments become clearer when we look a little more closely at a few of the stories of catching animals for Hagenbeck from this period.

The Early Traders

A sensible place to begin is with one of the earliest accounts of Hagenbeck animal catchers, the 1870s souvenir pamphlet *Carl Hagenbeck's Thier-Karawane aus Nubien* (Carl Hagenbeck's animal caravan from Nubia).[29] According to the pamphlet, "for years the most important animal dealer, Mr. Carl Hagenbeck in Hamburg, sent his travelers every year to Nubia and Sudan," in order to "take over and ship to Europe" animals caught by the "Nubian hunters." Hagenbeck's agents, we are told—and here we should count such people as Casanova, Cohn, and Menges—are chosen for their familiarity with the land and its "half-savage inhabitants" and their fluency in Arabic. The text continues, describing the animal station:

The camp of the animal trader is commonly situated near a river, is always very extensive, and forms a small village unto itself. It holds the straw huts for the Europeans and their black servants, areas for the captured animals which are fettered in the shade of thick trees, the stable for the horses, and the kitchen in which one or more slaves prepare food for the group. The whole thing is surrounded by a high, thick thorn hedge made from felled trees. A thornbush set up nightly closes the only gate to this camp. Through the nights fires burn to keep the wandering carnivores away, and rarely does a dark night pass in which a lion or leopard does not visit the lonely camp, roaring for hours at the inhabitants until they scare the cat away with thrown torches or a few shots into the air. During the day the station is governed by bustling activity, familiar natives arrive in a constant flow, often out of sheer curiosity, but mostly to offer their service as hunters and to obtain hunting horses.[30]

Once the native hunters have been supplied with horses, ropes, and other necessary materials, they leave the zareba to return weeks later with the animals. In general, according to the pamphlet, the herd animals are simply chased until the young fall behind, traps are set for leopards and other cats, and young elephants and rhinos are captured after the "mothers" have been killed. After their capture, the animals are gradually transported to the station, where negotiations begin. The pamphlet notes that transactions could often last "for several days before the trader [came] into possession of the animal through a sum of money and small gifts such as tobacco, sugar, soap, and (especially coveted) empty bottles. The captured animals [were] always quite exhausted, and a third to half soon die[d] from the stress of the hunt."[31]

After a stay of often several months, Hagenbeck's early "animal catchers" would begin the transport from the interior to the coast. Again, the example of the "Nubian caravan" described in the pamphlet gives us some idea of the difficulty of getting the animals to Europe. First, the size of these caravans is often startling. In addition to smaller animals, the pamphlet's caravan included three rhinoceroses, fourteen elephants, nine giraffes, fifteen antelopes, and seventeen ostriches. Moreover, Hagenbeck's agents contracted two to four men to walk with each elephant, three for each giraffe, and two for each antelope and ostrich. Those animals too small or dangerous to be walked to the coast—the young lions, leopards, birds, and the like—were packed in cages and carried by camels. Finally, a large herd of goats was brought along to provide milk for the young and meat for the carnivores. According to the pamphlet, the "Nubian caravan" took fifty-four days to reach the Red Sea from the interior, marching during the cool hours around sunset and dawn. Barely half the animals obtained in the interior survived to the port of Suakin. From there, they were transported by steamship to Suez, then by rail to Alexandria, by steamship to Trieste, and finally by rail to northern Europe. The pamphlet concludes by noting, "From all of this one can appreciate the trouble and expense that such a transport requires, and it is hoped therefore that the public might compensate Mr. Hagenbeck for his great cost and sacrifice through friendly attention to the animal transport and its black Nubian attendants and hunters."[32]

Of course, as the trade evolved in this early phase, a certain regularity set into the business. Soon it became less and less necessary for Casanova, Menges, Cohn, and the other traders from predominantly Germany, Greece, and Italy who were active in northeast Africa to travel deep into the interior to contract with catchers. As Josef Menges noted in an 1876 article on the animal trade in northeast Africa:

> The animal trade, as it is now handled by the Germans, has become strongly regulated through years of practice and operates in general as smoothly as any other business. In October and November when the unhealthy period following the rainy season comes to an end, the traders travel from Cairo or Alexandria to Suez where they catch the Egyptian steamship which travels twice a month to Suakin and Massawa. The trip to Suakin lasts between four and six days. Suakin is the beginning point of the great caravan routes which lead through Kassala and Gedaref to Khartoum and through Gallabat to Amhara. The trading caravans travel for sixteen days to get from Suakin to Kassala; the path is fairly well traveled, but it is naturally only passable with camels, water being available every two to three days. In Kassala the trader rents a home where he will be sought by the nomadic Arab peoples of the interior. They will bring him those animals they have already captured which they wish to sell and will enter into further binding agreements for other animals. The only Arabs who systematically engage in the capture of animals are the . . . Hamran who have their hunting grounds on the Setit (a lower fork of the Takkaze which flows from Abyssinia). The land between the Setit, Gasch, and Atbara rivers is extraordinarily rich in wildlife and therefore there is no shortage of the young and catchable animals.[33]

With the arrival of the trader, Menges notes, the hunt would be stepped up, with the catchers giving special attention to those animals especially coveted by the trader. The animals would then be delivered to Kassala, where they would be purchased by the trader, along with those animals that had fallen by chance into the hands of other peoples. Menges notes that although the increasingly programmatic character of the trade had made the European's task somewhat easier, if an especially rare animal was sought, the trader would usually have to go into the field to contract directly with indigenous catchers. If, on the other hand, the trader was interested in only the standard assortment of giraffes, elephants, buffalos, hyenas, lions, leopards, antelopes, ostriches, and other birds and small animals, he could reasonably expect to wait in the relative comfort of Kassala for the animals to be delivered.[34]

It is not simply a coincidence, of course, that the procedures described in the "Nubian caravan" pamphlet and in Menges' article echo the histories of other early colonial industries. Indeed, the two works capture fairly well that moment in the consolidation of power in colonial lands when Europeans increasingly bypassed coastal traders and chiefs to move through the interior collecting their trade goods. Nor is it surprising, therefore, that the methods employed by the animal catchers paralleled the practices in other, more firmly established industries. Still, as Endre Stiansen emphasized in his study of the ivory and gum arabic trades in the Sudan, clear generalizations about colonial trades are difficult.[35] In the case of gum arabic, European traders tended to travel throughout the region and purchase the gum wherever it was available, and in the case of ivory, trade tended to occur at more permanent stations, but different practices were employed in different areas, and local traditions and the trader's abilities and expertise all played a part in how and where a given transaction took place. It seems fairly likely, though, that the trading stations used by the Hagenbeck "catchers" were at the same time being used for other trades and that similar techniques were employed in all these businesses. Traders, for example, often utilized variations of the "trust system" suggested by the "Nubian caravan" pamphlet, whereby supplies and goods would be given in advance to the indigenous people in return for the promised future delivery of goods. Largely stemming from practical trade considerations in areas without established banking systems, this system's deeply embedded debt structure drove the market as debtors trading ivory, gum, oils, or, in this case, live animals sought both to maintain their valuable contacts with the traders and, often impossibly, to extricate themselves from debt. In any case, general rules for the animal trade, as for other colonial trades, are difficult to establish.[36]

Nevertheless, it remains clear that the kinds of trading systems used by Hagenbeck's catchers carried significant speculative risk and required substantial capital support to cover the transportation and maintenance of the catchers themselves, the establishment and provisioning of trading stations, possible prepayments to catchers, and the transport and insurance of the animals. In the cases of the semi-independent traders such as Casanova, Menges, and Cohn, most of that cost was borne by the traders themselves. By the turn of the century, however, Hagenbeck seems to have been funding completely a majority of the animal-catching expeditions supplying his establishment. The reason, of course, for his willingness to bear more financial exposure for the expeditions becomes clear when one considers the potential profits of successful catching expeditions. According to Menges, who published a list of purchase prices of elephants, giraffes, rhinoceroses, antelopes, and lions in Kassala and their sale price in Europe in 1876, an elephant, for example, could be purchased in Kassala for 80–400 marks and sold in Europe for 3,000–6,000 marks; a giraffe purchased for 80–200 marks could be sold for 2,000–3,000 marks; and a rhinoceros obtained for 160–400 marks could be sold for 6,000–12,000 marks.[37] Menges argues that these returns—in the case of the rhinoceros up to 7,500 percent—must be considered in light of the often extraordinary costs of getting the animals to Europe. The costs of such trips can be imagined, he notes, when we consider that to move the animals to Suakin on the coast one must obtain 50–100 camels at 24 marks each for the trip, 50–60 people to walk with the animals, and supplies of water and food for the animals and people. Shipping the animals, then, from Suakin to Suez could cost as much as 150 marks for each elephant or giraffe and 130 marks for each smaller animal, with animals in cages costing 34 marks per cubic meter. Each railroad wagon load of animals from Suez to Alexandria would cost 330 marks; and tariffs were as much as 8 percent of the animals' value. Add to all this the costs of loading and unloading the animals at every stage of the trip, of housing and feeding the animals along the way, of the required tips to move the whole process along, and of having a third, half, or even two-thirds of the animals die along the way, and, Menges argues, the markups do not seem so extreme.[38]

That said, for the purpose of getting a general idea of the potential profits involved in this trade, Menges' 1876 figures lead to some impressive, if not wholly unexpected, outcomes. Take, for example, an easily imagined shipment of one rhinoceros, five elephants, fourteen giraffes, twelve antelopes, and seven young lions:[39] the animals, according to Menges' figures, could be obtained in Kassala for as little as 1,976 marks or as much as 7,200 marks. If we take the larger figure and add twice that amount to cover all the incidental costs of shipping the animals to Europe (my estimates suggest the actual cost to be under 10,000 marks for these associated expenses), these animals will have cost Hagenbeck a maximum of 21,600 marks—an admittedly large sum of money. Nevertheless, for this shipment of thirty-nine animals, Hagenbeck, again using Menges' figures, could expect to receive at least 61,000 marks and as much as 136,800 marks. Even halving those numbers to allow for a 50 percent mortality rate yields acceptable profit margins. To be sure, some Hagenbeck expeditions, such as the first attempt to catch a pygmy hippopotamus in Liberia, came back with little to nothing. Even those expeditions, however, were turned to the advantage of the firm as word spread that Hagenbeck would be willing to go after anything—the story was that Hagenbeck could supply "anything from a white elephant to a flea" and that he was "always ready to try the untried, and attempt the impossible."[40] But those expeditions remain the exception. Hagenbeck's business relied on volume, and when we consider the sheer number of animals moved by the firm from the 1860s to 1880s, it becomes clear how Hagenbeck rose to such wealth and fame. According to an 1887 estimate by Hagenbeck, in the first twenty years of the company, "he had sold at least a thousand lions, three to four hundred tigers, six to seven hundred leopards, a thousand bears of different varieties (at one time he had forty-two at the same time), and around eight hundred hyenas. Some three hundred elephants had passed through his hands, of which sixty-three Indian and four African elephants had come in 1884 alone. He had sold seventeen rhinoceroses of the three Indian species and nine of the African, while one hundred and fifty giraffes, of which in 1874 alone he sold thirty-five, and six hundred antelopes of diverse species including the rarest, largest, and most beautiful, had been traded by the company." The account continues by noting that in the same period the company sold 180 Bactrian camels, 120 dromedaries, and 150 reindeer and that in one remarkable shipment of animals, some 600 snakes had been sent of which 374 arrived alive in Hamburg.[41]

It is precisely within the context of Hagenbeck's success in delivering animals to Europe that we must understand his decision in 1873 to send his youngest brother, twenty-year-old Dietrich, to Africa specifically to catch young hippopotamuses for the company. This decision marks an important moment in the history of the firm. Although by 1873 Hagenbeck's animal business had become the most important source for animals from Africa primarily through its stations in the Sudan, the deliveries were still generally based on whatever Hagenbeck's agents had managed to get their hands on, mostly through trade and purchase. By sending Dietrich into the field to catch a specific animal, and, in this case, one of the most valuable animals on the market, the firm attempted to gain more control over the types of animals delivered.[42] At the same time, however, Dietrich's appearance as a new kind of catcher highlights a moment in the development of colonial trade. Whereas Casanova and Menges had bypassed traditional middlemen to trade directly with the peoples of the interior, albeit on similar terms, thereby diminishing the power of the coastal elites and consolidating the control of colonial administrations, Dietrich, and those who followed, took the next logical step and assumed control over the catching process itself while using local peoples merely as labor.

In the end, however, Dietrich's attempts to bring home even a single hippopotamus were unsuccessful despite months of effort. In a letter of March 14, 1873, the young hunter noted that after four weeks of searching, having seen many hippos, including as many as twenty animals in a group, and having shot four adults, he had been forced to retire to Zanzibar to recover from fever.[43] A letter from the end of April from Dar es Salaam reports another month of unsuccessful effort, increased frustration with the "laziness" of the "natives," and unrelenting bad weather during a three-day pursuit of a female hippo and her baby.[44] Finally, in a long letter of June 25, Dietrich reports that he had managed to capture a young hippo but that the animal had apparently been bitten by a crocodile sometime earlier and after six days had died from the wound.[45] Dietrich himself died shortly thereafter from a virulent form of malaria.[46] As the bourgeois magazine *Die Gartenlaube* (The garden arbor) noted, he represented "unfortunately yet one more sacrifice in the long list of those who have not returned from Africa."[47]

Hunters Become Catchers

While Hagenbeck's success at taking increasing control over each stage in an animal's delivery to Europe or the United States—from the animal's capture to transport and eventual sale to a zoological garden or other collector—helped the company become the preeminent conduit for the larger exotic animals into the world market, more modest collections of animals put together by naturalists, colonial administrators, "great white hunters," and commercial hunters continued to be purchased by Hagenbeck and other traders. How these hunters and their hunts were portrayed varied, of course, according to the interests of collector himself, but a closer look at most of the stories about them will show, often in sharp contrast, the differences between these amateur catches and the type of professional catching which would become the hallmark of the Hagenbeck business by the turn of the century. Hans Dominik's elephant adventure, "Catching Elephants among the Mwelle," from his 1901 *Kamerun: Sechs Kriegs- und Friedensjahre in deutschen Tropen* (Cameroon: Six years of war and peace in the German tropics) presents a rather striking example. Prompted by a conversation in the mid-1890s with Ludwig Heck, the director of the Berlin Zoological Gardens, Dominik, a captain in the German colonial army in Cameroon, had decided to catch a baby elephant. Heck had told Dominik, who had already brought a variety of animals from the colony, that the day when the first elephant arrived in Berlin from German Africa would be a day of celebration.[48] Dominik concluded that his best chance of success would be if he captured an entire herd and then obtained the young by killing off the adults. Though bloody, Dominik's plan was practical, and his ability to execute the task was repeatedly noted with admiration by other writers.[49]

After spreading the word that the first people to lead him to a herd of elephants would collect "a large reward," Dominik finally received information in early October 1898 that Mwelle hunters had discovered a herd and were awaiting his instructions. After his arrival, Dominik observed the herd with his guides, Zampa and Amba: "There was little activity among the animals. The calls of the working humans which carried clearly through the quiet forest hardly appeared to bother them. One bull stood apart, preoccupied with tearing twigs off branches with his trunk and consuming the leaves. Closest to us stood a cow using her trunk to lovingly caress her baby, which was barely larger than a pig and stood between her legs. A few animals ate—sweeping together and ripping up low-growing grasses and using their trunks like sickles—most of them appeared to be sleeping. . . . We seemed so small, so insignificant when compared to the mighty animals in the mighty wild."[50] A fence was constructed during the night to thwart the elephants' escape, and the hunt began the next morning. In his account, Dominik begins by describing the scene when he and his guides first came across two adults: "One after the other, a head turning to the left to pull up something green from here and there, the elephants came slowly toward us. The safeties were released. 'You, the second,'

I whispered to Zampa. Now we had the animals ready. I fired at the right ear of the foremost animal. At the sharp crack, the elephant threw its trunk into the air and trumpeted loudly. The short tail stretched out far, he turned upon himself like a top. In this moment, Zampa had also fired. Close before me the second animal buckled at his knees, but quickly stood up again and followed the incessantly bellowing and bleeding lead bull which pushed up the hill." Firing after the animals and then tracking them over a small rise, Dominik found the two elephants beside each other:

> There lay one of the animals; apparently the spine had been hit because the elephant had only collapsed in the rear and was in a sitting position. Like columns, the forelegs projected from out of the ground, the head and trunk swung left and right: a muffled moan sounded, thick clumps of blood flowed at the side, a sign that the lungs were also wounded. The other stood next to him, motionless except for his trunk. He blew frequently, and with his trunk threw soil on himself. Our approach didn't seem to bother the animals. We crept around them. I had the eye of the sitting giant exactly in the rifle sight, when beside me Zampa fired. The standing elephant trumpeted loudly. Now I squeezed the trigger and the animal collapsed onto its side. The other elephant was still standing; finally with the first shot from my second chamber he collapsed. Close beside one another lay the two giants in a massive pool of blood. Amba and Balla were already there; with their sharp machetes they cut through the trunks, which were half the thickness of a man. The animals were still breathing. As if from a fountain, the red blood sprayed up from the thick arteries onto our clothes as we stood beside the animals examining our guns and discussed how we should proceed with the hunt.[51]

The fascination with grisly detail which permeates this narrative continues as the hunt progresses. Soon Dominik encountered a female with a young calf; after several shots, also graphically described, the female was dispatched with a shot in the left eye. The calf was roped and tied to a tree, where it "churned up the soil with its small tusks, bellowed and moaned, charged backwards, stood on its head, and foamed at the mouth in rage as bloodshot eyes protruded from its head."[52]

Three remaining calves were soon captured as well, one dying of suffocation after having its trunk pulled between its forelegs and tied to its rear legs so that it "breathed with difficulty and lay on the ground like a large gray sack."[53] Another calf died during the night of wounds sustained in the capture, but Dominik had still managed to secure two calves from the herd and soon added three more to his collection. Two died a month later, but the remaining three apparently thrived in their new environment, and one found its way through Hagenbeck to the Berlin Zoo, where it was seen by literally thousands of Berliners who lined up to view the newest acquisition from the colonies.[54]

In his 1887 biography of Hagenbeck, Heinrich Leutemann was frank about the destructiveness of animal catching. "For the animal trader," Leutemann noted, "the method of capture is, from a business point of view, a trivial issue. For him the most important thing is that the animals arrive healthy, that they are, therefore, well packed." As Leutemann made clear, the animal catcher's primary concern was not obeying the cultured rules of the hunt but doing his business as quickly and efficiently as possible. Thus, animal catching was rarely concerned with the ritualized confrontation of man and animal on a field of honor; rather, for quite practical reasons such as care and transport, the catcher generally ridded himself of the adults as simply as he could and concentrated on obtaining the young with as little impact on the animals as possible. As Leutemann put it:

> Without exception, lions are captured as cubs after the mother has been killed, the same happens with tigers, because these animals, when caught as adults in such things as traps and pits, are too powerful and untamable, and usually die while resisting. . . . The larger anthropoid apes can, in addition, also only be captured—taking into account occasional exceptions—quite young beside the killed mother. The same is the case with almost all animals; in the processes, for example, giraffes and antelopes, when hunted, simply abandon their young which have fallen behind, while in contrast the mother elephant more often defends her calf and therefore must be killed, as is the case with hippopotamuses. . . . Also in the case of the rhinoceros, the young are captured from the adults, which are usually killed as a result.[55]

The point is that, at least as far as its general approach to the efficient collecting of elephants, Dominik's tale seems generally predictable; we should not be surprised that this type of hunt was a relatively bloody affair. Or perhaps we should. In the end, what is absolutely basic about Dominik's tale, what marks it as a particular kind of hunting story in sharp contrast to the earlier stories of Menges and the works of Hagenbeck's professional catchers described below, is Dominik's decision to present, and indeed wallow in, all the gruesome details of his encounter with the elephants. His description, for example, of standing essentially in a blood fountain examining his gun after dispatching the first two elephants—so particularly preoccupied with and fascinated by blood and death—suggests that this story is only peripherally about catching elephants. Centrally, in fact, Dominik's story is about killing and the conspicuous display of power.

In this respect, Dominik's narrative is echoed by a whole range of "catching stories" from this period. A typical item in this genre, for example, is Hans Hermann Schomburgk's 1910 *Wild und Wilde im Herzen Afrikas* (Wildlife and savages in the heart of Africa), in which Schomburgk describes his life as a professional ivory hunter and later animal catcher for Hagenbeck. Having arrived in southern Africa in 1898 ostensibly to start a farm, and after several expeditions to the interior and a stint in the Natal Mounted Police, Schomburgk traveled north and, by the fall of 1906, had become a professional ivory hunter for largely pecuniary reasons. Fascinated by Africa and passionate about being out in the bush, the young explorer nevertheless needed a livelihood, and hunting elephants provided sufficient income to fund all manner of enthusiasms for exploration, sport, and natural history (fig. 16). To be sure, in his stories of his adventures—like those of the other figures in this chapter—Schomburgk creates his own character. Here he presents himself as the daring and adventurous young man, a character that is, of course, an exaggeration. In the introduction to his *Wild und Wilde,* however, Schomburgk is sensitive to possible criticism about the large number of elephants he in fact killed and the many more he was reputed to have slain. In response, he notes succinctly, "In my travels I have received financial support from no one, and if I should be accused of having killed many elephants, I must acknowledge myself as guilty. At the same time, though, I can console myself that I always hunted according to the rules and with very few exceptions—where there was mortal danger—did not bag any cow elephants."[56] In point of fact, the first two elephants Schomburgk shot—in a situation that was apparently far from mortally dangerous—were females, but the author glosses over this detail and jumps ahead: "It is impossible for me to describe just how proud I was to sit myself down on the first elephants I had slain by myself."[57] In any case, after hunting ivory for a year and a half, Schomburgk took a lead from Carl Schillings, who had tried several times to catch and keep alive a baby elephant in German East Africa. After some failed efforts, Schomburgk eventually found himself favorably situated to catch the perfect calf. On April 21, 1908, he sighted a small herd and discovered among the cows a suitably young specimen. After separating the cow and her calf from

Fig. 16. Hans Schomburgk as an elephant hunter. From *Wild und Wilde im Herzen Afrikas* (Berlin: Fleischel, 1910). © Hans Schomburgk—Archive Jutta Niemann.

the rest of the group, the ambitious hunter tracked the animals for three days before, quite unexpectedly, he came across the pair in high grass. Immediately the mother cow—"a huge tuskless animal"—charged the hunter, who "nevertheless luckily succeeded in landing a shot at the base of the animal's trunk."[58] Schomburgk continues: "She collapsed in the rear and gave me the opportunity to jump quickly sideways and bring to bear a deadly shot, after which she immediately died. Obeying the laws of nature, the young animal remained standing beside its mother. . . . Until my men arrived, I observed how the pitiful little baby continuously ran about its mother while hitting her with his trunk as if he wanted to wake her and make their escape."[59]

In his reminiscences, Schomburgk included a photo of "Jumbo" standing in front of his dead mother with the caption "Obeying the laws of nature, the young animal remained standing beside its mother" (fig. 17).[60] Killing the mother was, of course, only the beginning of Schomburgk's task in catching Jumbo, but soon the animal was under control, and before long it had formed an attachment to the hunter and his men and was eating well, eventually taking milk from a bottle. In mid-August, some four months after the elephant's capture, Schomburgk began the trip to the coast. The walk was slow because the young animal's feet often became sore on the march, necessitating that the caravan stay, for example, for eleven days in Tringa. By mid-September, though, Schomburgk's party had reached Morogoro, where both he and Jumbo were a sensation. From there the party took a train to Dar es Salaam, where Jumbo gained fame as a sandwich-sign carrier announcing upcoming concerts and other newsworthy events. Despite Ludwig Heck's suggestion to Schillings that the day the first German East African elephant came to Berlin would be a day of celebration, Heck, according to Schomburgk, did not respond to cables offering the animal for sale. Instead, one of Hagenbeck's collectors—Christoph Schulz—saw the animal, bought it, and immediately sent it to Stellingen.[61] Schomburgk reported, "On April 5, 1910, I was able to greet Jumbo back in Germany, and he now lives in Carl Hagenbeck's Animal Park where I visit him often." "That is how Europe," Schomburgk declared, "came into possession of its first East-African elephant."[62]

The modest but critical point to the stories of Dominik and Schomburgk is that these hunting tales are primarily about killing, about the ostentatious af-

Fig. 17. Hans Schomburgk: "The elephant Jumbo beside his dead mother." From *Wild und Wilde im Herzen Afrikas* (Berlin: Fleischel, 1910). © Hans Schomburgk—Archive Jutta Niemann.

fectation of "courage," about sitting on top of dead elephants and standing in pools of their blood. In describing his career, for example, Schomburgk writes: "No matter what people believe, it nevertheless takes courage to do battle with an outraged bull elephant in the thick brush or high grass. Three times I've been under an elephant; once I was thrown in the air by an elephant; an elephant trampled and ripped apart one of my carriers, and if I have managed to come out of all of these situations with a relatively intact skin, I have been lucky."[63] Fond of writing about the lack of sentimentality in nature and about how, in nature, the death of an individual is meaningless and only the survival of the herd significant,[64] Schomburgk found repeated occasions in his many books to praise the unappreciated courage of the great hunters who risked their lives in pursuit of the most dangerous quarry.[65]

It is precisely Schomburgk's early style of imagery and writing which would become a sheer impossibility in later accounts of animal catching.[66] An image like that of Schomburgk perched atop a bicycle on top of a dead elephant would simply be unthinkable for later catchers (fig. 18)—and indeed, after only a few years, unthinkable for Schomburgk himself. In fact, a general sense that "un-

Fig. 18. Hans Schomburgk on a bicycle. From *Wild und Wilde im Herzen Afrikas* (Berlin: Fleischel, 1910). © Hans Schomburgk—Archive Jutta Niemann.

necessary," "thoughtless," or "disrespectful" killing must be avoided in descriptions of animal catching became more and more clear to everyone in the business. During negotiations to procure a young rhino for the Bronx Zoo in 1902, for example, then director William Hornaday asked Hagenbeck to be circumspect in talking about how the animals were obtained. Hornaday writes:

> I have been greatly interested in the fact that your letter gives me regarding the capture of the rhinoceroses; but we must keep very still about forty large Indian rhinoceroses being killed in capturing the four young ones. If that should get into the newspapers, either here or in London, there would be things published in condemnation of the whole business of capturing wild animals for exhibition. There are now a good many cranks who are so terribly sentimental that they affect to believe that it is wrong to capture wild creatures and exhibit them,—even for the benefit of millions of people. For my part, I think that while the loss of the large Indian rhinoceroses is greatly to be deplored, yet, in my opinion, the three young ones that survive will be of more benefit to the world at large than would the forty rhinoceroses running wild in the jungles of Nepal, and seen only at rare intervals by a few ignorant natives.[67]

Hornaday's letter to Hagenbeck clearly suggests a certain sensitivity to discussing the unfortunate slaughter of animals in the process of catching them, in contrast to the tales of Dominik and Schomburgk; nevertheless, the two approaches to discussing animal catching have much in common. Both approaches, for example, essentially conflate the animals, people, and larger environments of territorial possessions into generic objects valued specifically for their importance to the colonial powers. Thus, Hornaday, in a justification for animal catching which is common enough even today, argues that the captured rhinoceroses will far more "benefit the world" than forty adults "running wild in the jungles . . . seen only at rare intervals by a few ignorant natives."

Sentiments such as these helped drive the often brutal nature of this business and, moreover, the often almost despotic behavior of the catchers themselves. Referring to the peoples of Cameroon whom he was charged with subduing, Dominik, for example, declared, "They must know that I am their master and that I am the stronger, and as long as they do not believe that, they must be made to feel it, and I mean severely and pitilessly, so that rebellion will pass by them for all time."[68] Typically, in signing peace treaties with Dominik, "freed" villages would be required to pay tribute in goods and men, with the men being taken away chained for carrying and plantation work. Dominik's deliveries of hundreds of forced laborers to plantations and factories over the course of his years in Cameroon seem rather typical in this regard, as do his use of hostages to obtain carriers and his whipping and shooting of disobedient workers.[69] In this environment, beating and chaining workers was not considered particularly exceptional or barbaric behavior but appropriate and necessary in order to get the "natives" to work. As one Cameroon planter succinctly responded to critics at home in Germany in 1904, "Our people must not be allowed to forget that the politics of colonialism is the politics of conquest, that we are invaders in order that the land be brought to culture and developed for the German nation. We take it from the natives to exploit it for our purposes. We force the natives to work for the cultural development of the colony. This work must be demanded from them otherwise the colony will remain at the status quo at which we took it over."[70]

What we must recognize is that however romantic or adventuresome the exploits of the animal catchers of this period may have been represented as being, in most cases this business had a ruthless side in its dealings with animals, people, and environments which was almost a necessary component of the trade. The animal trade was a labor-intensive, extractive enterprise that, like most colonial industries (including tobacco, coffee, cocoa, saltpeter, rubber, ivory, tropical woods, palm nuts, and palm oil), was often based on forced labor and was highly destructive in its use of land. Dominik is not, in this regard, somehow unusual. To be sure, with readers at home increasingly uncomfortable with the use of force against their colonized peoples,[71] few authors in their published adventures of catching animals dwelled on their dealings with indigenous people beyond emphasizing their own abilities to work closely with "lazy," "uncooperative," "savages."[72] Nevertheless, descriptions of actual beatings of carriers and other laborers are present in these accounts. In his story of catching pygmy hippopotamuses for Carl Hagenbeck in 1911–12— reproduced in the *New York Times* as well as in the *Bulletin* of the New York Zoological Society—Schomburgk, for example, refers repeatedly to beating and whipping his workers and even holding a "chief" at gunpoint to secure additional carriers for his precious cargo.[73] Early on in the story, he writes, "I thought the time had come to teach my carriers a lesson. We were too far from civilisation already to fear desertion. When I called the boys in the morning to start, nobody came, so I called up my headman and asked him very quietly if the boys were packing up. 'No,' was the reply, 'they do not want to start yet.'

Without saying another word I took up my Browning and put 7 shots through the roof of the boys' hut.—And they came quickly! From this minute I took up the reins and after I had picked out the biggest and laziest of the motley crowd and had given him personally a good hiding, I had no further troubles." "Troubles" followed, however, in his second expedition into the interior in pursuit of pygmy hippos, during which at one point, Schomburgk claimed, he faced "open rebellion," to which he responded as follows: "I slipped the Browning into my pocket, took my hunting crop and went among them. Clash, crack went the whip on the naked body. A few straight hits from the shoulder on the jaws of those who did not move and quicker than I can tell it I drove the mutinous crowd before me like a herd of sheep! The result of the rebellion for the boys was that I stopped their rations for three days and the allowance of gin for a month."[74]

Nevertheless, just as in this period the often callous aspects of colonial rule began to draw criticism at home, so, too, did the often wasteful and "inhumane" methods of the animal catchers working within colonial regimes. On the one hand, those criticisms took form in the "sentimental" condemnations issuing from Hornaday's "cranks," the same type of sentimentalization of animals driving such diverse cultural phenomena as the Steiff stuffed toy animal business and the "nature faker" controversy surrounding highly sentimentalized nature writing at the turn of the century.[75] Other critics soon came forward, however, from often surprising quarters. Richard Meinertzhagen, for example, a particularly destructive, if also complicated, sport hunter, naturalist, and member of the British colonial army in Kenya in the first decade of the twentieth century, could not accept what he judged as the lack of sportsmanship in catching animals for zoological gardens. Arguing that the commercialization of hunting was responsible for the devastating declines in animal populations in Africa, Meinertzhagen believed that perhaps the most horrible of those commercial enterprises was the trade in living exotic animals. "I wholeheartedly disapprove of the catching up of wild animals for zoological gardens and condemning them to solitary confinement for life amid squalid surroundings," Meinertzhagen wrote. "I have seen a good deal of the 'catching-up' stage and can vouch for the high rate of mortality, for the shooting of mothers to obtain the young, for deaths in transit and for the cold-blooded nature of the catchers-up."[76] Even figures within the world of zoos began to take a stand against what was seen as the pernicious effects of the commercialization of animal catching. Friedrich Knauer, for example, the director of the zoological gardens in Vienna, admonished his colleagues shortly before World War I: "It is clear that the high prices which are paid for rarities have spurred the professional animal catchers and dealers to extreme efforts. Were the demand from zoological gardens for living musk oxen . . . not so brisk, it would hardly have occurred to the Norwegian fishermen to search the east coast of Greenland and slaughter whole herds of this, until recently, rarely endangered form of cattle. 28 musk oxen had to die so that the Danish expedition sent to Greenland in 1900 . . . could get possession of a young live musk ox for the Copenhagen Zoological Garden. . . . And so it is with many other rare species—we need to think only of the white-tailed gnu, of the varied representatives of the ancient Australian fauna, of the giant tortoises of the Galapagos Islands, of the birds of paradise."[77] Partly in response to the growing criticism of the animal-catching business and partly as a consequence of the careful refinement of how Hagenbeck wished his company to be perceived, heroic stories such as Dominik's and Schomburgk's were gradually displaced by narratives highlighting the more "civilized" hunting practices of Hagenbeck's quite new professional catchers such as Christoph Schulz.

The Professional Catchers

Of the many tales written by or about the Hagenbeck animal catchers, perhaps that which best marks the first decades of the twentieth century is Christoph Schulz's 1922 *Auf Großtierfang für Hagenbeck: Selbsterlebtes aus afrikanischer Wildnis* (Catching big game for Hagenbeck: Personal experiences from the African bush). Based on his experiences in German East Africa in the years immediately preceding World War I, the book went into several editions and was also abridged into a more popular form.[78] In the book's introduction, Eduard Elven, who met Schulz in Africa before the war and later found himself a fellow prisoner of war with the animal catcher in Malta, suggests the book's main appeal. Comparing it with the broader literature about hunting animals in East Africa, Elven concluded that the book is unique because, although the chase is an important element, keeping animals alive is Schulz's main task. Consequently, Elven noted, the book focused on "the sufferings and joys of an animal catcher, his life on the steppes and during the transport of the animals, as well as the fashion in which he is able, in the middle of the wilderness, to catch the animals which we admire as the ornaments of our zoological gardens."[79]

For those familiar with hunting literature from the period, the book presents few surprises; essentially, it is an edited field journal. The book has no plot; rather, Schulz, who was often photographed working alongside his wife, Elisabeth (fig. 19), adopted a typical approach for the genre and organized his field notes into a series of adventures divided by animal type—the hunt for buffalo, for giraffe, for zebra, for lion. The detail of geographical information is extensive, while the observations of animal life are told in an unstudied and matter-of-fact style that demonstrates throughout a primary focus on keeping the captured animals alive. By and large, *Auf Großtierfang für Hagenbeck,* which shared many features with highly popular works of travel literature in the period, transported readers to a land both strange and familiar. Filled with "exotic" animals and people, and describing a place where ordinary Germans could build an almost ordinary life, the book was generally more concerned with detail and problem solving than with drama.[80]

Thus, unlike the heroic hunter-catchers such as Hans Schomburgk who somehow always managed to bring the deadly shot to bear on the charging animal in the final moment, and also unlike the tales of Casanova and the "Nubian caravan," which posited animal-collecting stories as a series of exciting and deadly adventures with savage peoples and arduous tasks, Schulz is simply not particularly interested in describing his life as filled with danger and death. Whereas Schomburgk, the professional hunter, illustrates the story of his famous catch of a young East African elephant with a photograph of "Jumbo" beside his dead mother (fig. 17), Schulz highlights the presence of his wife, who helped tend the animals at the Hagenbeck-Schulz farm on the foothills of Mount Meru in German East Africa. Indeed, more than a farm: Schulz and Hagenbeck, as the latter explained in a letter to William Hornaday about Hornaday's newly published *Our Vanishing Wildlife,* began to think of this property in Africa as a reservation. Hagenbeck writes in English: "You are quite right about the destroyment there is going on with all kinds of animals and birds. And not alone that I fight against that—I get more and more friends which assist me, and I have brought it so far for our German East-African districts that we have worked out very severe restrictions. It will interest you that I have got from the Government more than 100,000 acres of land as a wild animal reservation."[81] Schulz corrals adult zebras and lassos young giraffes. On

Fig. 19. Christoph and Elisabeth Schulz, ca. 1930–35. Courtesy, Schulz Collection, Lampasas, Texas.

his farm, he and his wife settle down to rear animals, care for them, and ship them to Europe. While he is out surveying, his wife, an altogether new presence in a work such as this, feeds and tends the tame animals at home. Schulz writes that he found "great assistance in the care of the animals in [his] wife who, during [his] absence, faithfully cared for [their] charges and never shied from inconvenience or work." He continued: "She is in large measure to be thanked for the great success I had in the rearing of the animals, and for the fact that I was able to bring back alive and healthy the first giraffes, oryxes, elands, aardvarks, and many other representatives of the East African fauna to Germany."[82] When Schulz sets out after a baby rhinoceros, he attempts to do so without killing the mother by separating the young from the adult with the use of trained dogs. In catching hippopotamuses, he does not shoot the adults but builds instead special pits designed to catch only the young. Not surprisingly, Schulz has little affection for the thrills and joys of more traditional hunting. "For me," he writes, "there is no greater pleasure than to observe wildlife in the wilderness, to obtain young animals alive and unhurt, and to take care of them, so that they can be kept as long as possible in the zoological gardens where they will serve the advancement of science and the popular education."[83]

The general impression that one gains in reading works such as that of Christoph Schulz is that while animal catching may at one time have been a wasteful, perhaps even cruel profession in which Europeans sought little else than the maximization of profit, enlightened catchers and dealers now control the trade. These new catchers, we are told, truly understand the needs of animals; indeed, these new men, and women, cherish animals and catch them largely because the public back home needed to be shown just how wonderful the creatures of the world are. They hunt not for mere sport but because a captive live animal can show the world the need to treasure and protect all animals. Of course, by the time Schulz went to Africa to work for Carl Hagenbeck in 1908, the company had already come a long way in defining itself as an organization primarily motivated by a love of animals. Indeed, as will become clearer later in this book, Hagenbeck's Animal Park in Germany gained something of an echo in the Hagenbeck-Schulz African "farm" or "animal reservation": they were both quickly cast as places where animals could live in safety, protected from the predations of humans and removed from the fight for survival by the kindly Hagenbeck.

Schulz and Hagenbeck sought to reconstruct themselves and their professions as true friends of animals, utilizing throughout references to a beneficent and ameliorative process of progressive civilization, as well as metaphors of and literal references to "families" of animals and people. Not surprisingly, in his memorial in the Animal Park, then, Hagenbeck is portrayed standing with a calm hand resting on his favorite lioness, "Triest," who, the story goes, threw herself into battle against a suddenly dangerous group of cats in order to protect her keeper during one of his customary visits to the lions' den. As Ludwig Zukowsky, one of Hagenbeck's assistants, noted, "It was touching when [Hagenbeck] spoke of his lioness 'Triest' who was devoted to him like a dog. . . . Never did Hagenbeck appear greater among his animals as when the royal lioness lay overcome at his feet, not as a slave, but as a friend beside a friend."[84] At Hagenbeck's grave, "Triest" is seen again lying across her master's tomb, a gigantic representation of fidelity and love.

The basic problem with this new image put forward by Hagenbeck, however, is that it remained difficult for the company to portray itself as a protector of animals when its business rested on the capture and selling of whatever came into its hands. This is the point made in a cartoon of 1893 showing a wagon jolting its way through a jungle as two ostriches stick their heads in the ground and

Fig. 20. Hans Oberländer, "Hagenbeck Is Coming!" *Die Fliegende Blätter,* 1893.

all manner of animals flee before it screaming "Hagenbeck Is Coming!" (fig. 20). The cartoon was meant to be humorous, and, with the caricatured animals, most people today would probably agree that the illustration is still funny. At the same time, the satiric undertone of the picture—its playful reversing of the Noah's Ark motif—would have escaped few, and it was to this type of critique that Hagenbeck and his catchers began to respond. Indeed, throughout the world of Hagenbeck in the 1890s and first decades of the twentieth century, we see the company wrestling with a public image complicated by the very nature of its business. This struggle is apparent, for example, in the popular images of Hagenbeck and his walruses (fig. 21)—images that have been brought forth repeatedly over the decades to demonstrate Hagenbeck's love for his animals. The photographs present an elderly Hagenbeck surrounded by his charges, who mob him for special treats as a crowd of onlookers observe with fascination and delight. As one of the park's scientific assistants recalled, "Carl Hagenbeck could occupy hours with these enchanting walrus babies and the Park's visitors have often been able to see with their own eyes what a deep love of animals lay hidden in the Hanseatic businessman. Over and again he stroked the rough hide of the little ones with a loving hand, praised them, and gave them treats, while the thick bodies affectionately snuggled up close to him."[85]

In his own memoirs, Hagenbeck briefly explained the method used to capture the young animals: "In order to catch the young, it is usually necessary to kill the mother. One of the animals to be found at Stellingen was seized this way:

Fig. 21. Carl Hagenbeck and the walruses at Hagenbeck's Animal Park, ca. 1908. *Von Tieren und Menschen: Erlebnisse und Erfahrungen* (Leipzig: Paul List, 1908). Courtesy of Hagenbeck's Tierpark.

the killed mother was pulled up close to the boat and all maintained a complete silence, until the young one came and climbed upon its dead mother's back. At that point it was, naturally, not difficult to overcome the unassisted young animal." In the end, according to Hagenbeck, "with the capture of the five young ones recently brought to Stellingen, sixty-eight other walruses were killed."[86] Even in this case, in which the captive walruses were essentially a by-product of the commercial walrus hunt—the sixty-eight walruses were clearly not killed in order to catch the young—the company struggled to dissociate itself from the forces of commerce, power politics, and death at the heart of its business.

The End of the Heroic Catcher

All these points are driven home in one of the more remarkable works about a Hagenbeck animal catcher, Wilhelm Munnecke's 1931 *Mit Hagenbeck im Dschungel* (With Hagenbeck in the jungle). At its most basic, Munnecke's work poses as an account of the life and adventures of Carl Hagenbeck's half brother John, whose portrait for the volume shows a man comfortable in business clothes and pince-nez glasses. In 1884, at about the age of eighteen, John Hagenbeck set out on his first trip to the East to organize an exhibition of indigenous people for his brother. More trips followed, and by the end of the 1880s, John had settled in Colombo, where, by the beginning of World War I, he had established himself as a successful entrepreneur whose wealth stemmed from coconut, rubber, and cocoa plantations, from partnership in a number of short-lived companies, and also largely from catching animals in India, Sumatra, Java, and Ceylon.[87] John Hagenbeck published several volumes of his travels and adventures, and more were issued after his death.[88] Munnecke's work, however, published in 1931 but focusing on the years before World War I, though presenting itself as an account of the famous animal hunter, is at least partly fictional. Barely suggesting a timeline (Munnecke brings forth a handful of dates that are usually vague and often inaccurate)[89] and depicting main characters and events that seem merely to be modeled after real people and occurrences, the book resembles more a historical novel than a memoir or record of someone's experiences in the wild—the staple form of the hunting chronicle.

Munnecke begins his work in the suffocating stillness of a midday jungle heat in which all life rests in the shade. "Not a blade of the thick, man-high grass," he writes, "shows a sign of life. Long, slender, stands of bamboo rise menacingly like lances into the heights, protecting in an endless phalanx the hulking muddy primeval forest."[90] In this stillness, a bull elephant is banished from his herd for contesting the dominance of the lead male. Now, as he crashes through stands of thick bamboo, he becomes the "rogue" Munnecke describes as the greatest fear of the jungle. The elephant crashes through villages where the "huts of the natives rest peacefully on stilts under protecting palms" and where "children play naked with shaggy dogs" and "nothing disturbs the peace."[91] Children and huts are left crushed in the wake of the terror. In the end, the only force that can stop the mad elephant is an unnamed European—John Hagenbeck—and his gun. Munnecke describes the eventual confrontation:

> With an unholy, threateningly-raised trunk, the Rogue rages out of the primeval forest and stands still in the middle of the path. . . .
>
> Bang! resounds a shot.
>
> The hollow echo is still making its rounds when the giant sinks forward on its knees. Breathing laboriously and hoarsely, he thrusts his powerful tusks into the ground, steadies for a few seconds and then pushes himself up again with his hind legs and tusks.
>
> Then a second shot cracks into the massive skull slightly behind the ear, and silently the colossus falls on its side.[92]

This "dethronement" of the "tyrant of the jungle" foregrounds recurring themes of Munnecke's work. Telling the stories through a disembodied voice able to follow animals into the jungle and describe their world through their eyes, the author focuses on the dramas in the lives of jungle animals and the place of the European hunter in those dramas. In the stories of Munnecke's *Mit Hagenbeck,* there is a pervading sense that things are rarely what they seem, that violence and death are always close at hand, and that the European is somehow both the deliverer and undoing of the jungle.

After the opening story of the elephant, Munnecke finally introduces us to John Hagenbeck as simply "the big European" who has followed his trackers to the den of a tigress. "Don't shoot, whatever happens don't shoot!" are the first words spoken in the book by Hagenbeck as the party observes the mother tigress leaving her lair. "Without reason," he warns, "one does not kill."[93] When the members of the party are assured that the mother tigress is no longer in the vicinity, they approach the den, and Hagenbeck sends his tracker in after the cubs: "Quietly he creeps deeper into the gradually widening passage. A lost, dull light breaks from above into the dark hole and plays on the floor in small moving specks. Quickly he sets to work, rushes outside with his booty, turns it over to the others, and returns back just as quickly. Yet another time he scurries through the dark passage, and yet one more time he grabs the whimpering, squirming golden flecks. Then the nest is empty."[94] With the cubs tucked into the hunters' clothes, the party hurries back to its camp. Hagenbeck requests milk from a nearby village for his new charges, the hunters eat happily after their day's labor, and all retire for the night. Munnecke's text, however, returns to the tigress: "There, above in the cave, however, a mother moaned complainingly; a mother whose three children had been abducted by the cruel hands of humans. By humans against whom she had never done anything, whose path she had always stepped aside from as soon as her nose caught wind of them. By humans, who, with malicious, refined, and ingenious weapons, attack, torment, jail, abduct, and in cowardly fashion kill the animal from behind."[95] After realizing what has happened to her "children," the mother tigress sets off in search of vengeance. Eventually, after a night spent watching the fires of humans, she begins a killing rampage against the natives. Then, still moved by her wrath, she attacks and kills the stabled horse of a Dutch surveyor who is erecting a telephone wire through the jungle. Again alluding to the force of "civilization" creeping into the natural world, Munnecke describes how this surveyor not only "pulls the long thin wire which at every moment connects him out of the depths of the jungle with civilization"[96] but also shoots the "the mother" the next night when she returns to feed on the horse. The following day, the tigress's corpse is brought to a village suspended on bamboo poles to the joy and relief of all. Meanwhile, Hagenbeck rears the young tigers to become performers for circuses.[97]

The unease with which Munnecke describes the death of the "mother tigress" is picked up repeatedly in *Mit Hagenbeck* as we watch the anguish and confusion of animals caught by Hagenbeck to be brought to Europe and, indeed, as we watch the slow and painful death of the jungle itself. Although the book is ostensibly intended to tell the heroic story of the animal catcher John Hagenbeck, it seems to be more about the death and suffering of animals. If the tone of the first half of the book, for example, is set by the story of an elephant gone mad who must be killed by Hagenbeck, the second half of the work—which focuses on Hagenbeck's adventures in Ceylon—begins with the story of another elephant, this one an ancient member of a herd dying a prolonged and agonizing death outside the view of humans.

For weeks the ancient elephant had been trailing its herd, its withered trunk hanging to the ground while its five-foot-long tusks fatigued the huge head. Occasionally a sign of life could be seen in the ancient animal's trunk, but then it would soon fall lifeless once again, and though the old bull tries desperately to grasp at leaves with his mouth, he becomes weaker and hungrier as the days pass. "Hunger gradually overcomes the Old One," the narrator explains. "He does not get angry, he does not complain. All he wants is shade and cool. Protection from the scorching, draining sun." Not even a small pool can ease his pain, however, for he is unable to use his all but dead trunk to drink. "Alone he carries himself along the jungle path. Abandoned by his herd, abandoned by his huge strength—the fate of the old."

One evening, however, a cool breeze from a lake reaches the elephant, who is standing quietly in the woods, and he breathes deeply. Knowing that a lake is near, the elephant begins to walk slowly to the water. "Through the thick hedge by the water," the narrator relates, "he forces his massive body, wades ponderously—his wide feet sink in the sludge at the bottom—up to his mouth in the water: water! Finally water!—And as if in a mating prayer, he closes his eyes reverently and drinks—drinks long and heavily." After taking his fill, however, the old elephant does not leave the water but stands quietly, listening to "sweet, loving, tempting" voices. The old animal's life slowly ebbs:

> Life becomes easy and dreamlike for him. Herds, vast elephant herds roll by him in waves. He runs with them, falls in battle, for one more time hot blood courses through his trunk and he swings it out of the water—and once more a surge of wind streams through it.
>
> With the fleeing of his life, he trumpets out in a roar his last breath.
>
> Then he sinks into the water.[98]

Munnecke's stories are deeply imbued with a romantic foreboding, a sense that the essence of the jungle is embraced by tragedy and that the "civilization" brought to the primitive forest carries an almost unavoidable destructiveness. Thus, while the Indian jungle is in some ways liberated by the power of John Hagenbeck (when he kills the rogue, for example), that jungle, as told in the story of the old, dying elephant, is itself fading away. *Mit Hagenbeck im Dschungel* is filled with animals who die in the wake of civilization but whose deaths are never seen as heroic or somehow just. In an adventure to catch orangutans, for example, two adults are caught in a frenzy of fear and distress, but their "baby lay crushed under a branch," and the two survivors soon die of sorrow.[99] In a sense, Munnecke's hero represents the last stage in the development of Hagenbeck's animal catchers. From Casanova and Menges through Schomburgk and Schulz to the fictionalized "John Hagenbeck," we can trace the evolution of the modern animal catcher. By the time we reach Munnecke's "hero," the animal catcher's task has become increasingly precarious. The focus of accounts about animal catching has moved away from the catcher to the animals themselves, and those animals are caught as much by the needs of civilization, science, education, and recreation as by some heroic catcher.

And what of Red Peter, the ape of Kafka's short story? Having been shot during the capture, he did not regain consciousness until he awoke between decks on the "Hagenbeck steamer." He explains: "I belong to the Gold Coast. For the story of my capture I must depend on the evidence of others. A hunting expedition sent out by the firm of Hagenbeck—by the way, I have drunk many a bottle of good red wine with the leader of that expedition—had taken up its position in the bushes by the shore when I came down for a drink at evening among a troop of apes. They shot at us; I was the only one that was hit; I was hit in two places." Red Peter was hit, he tells us, once in the face, leaving the red scar that led to his name, and once "below the hip." Although he acknowledges that he has been criticized in the press for doing so, Red Peter reports that he enjoys showing his public where the second shot hit him, insisting that nothing untoward is to be seen but "the scar made by a wanton shot."[100] Red Peter's desire to show his audience where he was hit by the second shot alerts us to a crucial point in his analysis of his capture: as brutal as the actual event may have been, and as brutal as his "middle passage" across the Atlantic was, it is, in the end, his exhibition that makes sense of his capture. Thus, Red Peter's critique of his capture turns necessarily on his critique of his life as an exhibited object, as an artist, among the civilized. As we shall see, if the expression "Hagenbeck Is Coming!" meant one thing in the far reaches of the globe where Hagenbeck's animal catchers plied their trade, the expression meant something else entirely in the places where Red Peter and others would eventually perform.

Although the name Hagenbeck first became known among animal catchers, buyers, and other dealers, the name soon spread into the popular imagination of people around the world who flocked to see the results of his catches—the exotic animals and, initially more famous, the exotic peoples brought for exhibit in Europe. Indeed, despite the fact that the Hagenbeck exotic animal business had been fully operating and open to the public since the early 1850s, it was the traveling exhibitions of indigenous peoples that made Hagenbeck famous beyond the limited environs of Hamburg. The Hagenbeck people shows, which began in the mid-1870s, made "Hagenbeck" a household name—a name that would eventually become more and more known as Hagenbeck applied what he had first learned from exhibiting people to his later exhibits of animals. In retrospect, the exhibitions of people seem to perplex most modern observers. Some people find them deeply disturbing, others find them somehow humorous, and still others seem to feel they are just one of those quirky phenomena of the late nineteenth century which appear both dark and naively quaint. For its part, the Hagenbeck firm appears to be alternately either mildly embarrassed or insistent that the public both understand the historical context in which the exhibits flourished and realize that the exhibits left little, if any, disturbing legacy. As we shall see, the legacy of these exhibits has been profoundly important in ways that have never been fully appreciated.

CHAPTER 3

"Fabulous Animals"

Showing People

⌒

"A True Copy of Life in Nature"

In his memoirs, Carl Hagenbeck noted that by the mid-1870s, his animal business had begun to slow considerably. Although the number and size of zoological gardens increased almost yearly, he argued that the acceleration of competition from more and more dealers (which both drove up the cost of obtaining animals and increased the number of animals available in the retail market) coupled with the growing success of the zoological gardens in keeping their stocks alive created an "overproduction" of animals and rapidly deflating prices. Hagenbeck's analysis of the problems undermining his business does not take into account the more general deflationary trends in Germany and Europe from the mid-1870s through the 1890s, but there seems little reason to argue with his basic observation that he was often stuck with difficult-to-unload assets while they literally ate up other assets. As his 1896 biographer, Wilhelm Fischer, put the problem: "Just when Carl Hagenbeck had succeeded, with diligence, energy, and a calm circumspection, in coming to dominate the African animal trade, happenstance and external conditions almost drove the business to insolvency. As the animals became more and more expensive for the dealer to acquire, their price in Europe collapsed to a degree such that, given the extreme risk entailed precisely in the animal business, even the coolest heads would have hesitated. Hagenbeck was forced, as he himself told me, to search for new sources of income to lessen his enormous losses so that he could 'struggle through with honor.'"[1]

According to the story told by the company now for more than a century, in the context of this awkward financial dilemma, Hagenbeck's old friend Heinrich Leutemann hit upon the idea that saved the firm. Having heard that Carl intended to import some thirty reindeer to meet the requests of a number of zoological gardens, Leutemann suggested that "it would certainly excite significant interest if the reindeer were accompanied by a family of Laplanders, who naturally would also bring their tents, weapons, sleds, and complete households along."[2] To the son of a traditional booth operator at the Hamburg Christmas fair, the idea must have struck a chord, and Hagenbeck quickly contacted his Norwegian agent. In mid-September 1875, a group of Sami,[3] with all their belongings and a herd of reindeer, arrived by ship in Hamburg. Hagenbeck recorded that as soon as he saw the group he knew the enterprise would succeed. "The caravan," he noted, "was comprised of six people and made a most striking impression. On the deck, the three male members of the troupe—small yellow-brown, fur-clad people—strutted about beside their reindeer. On a lower deck, however, we were offered a simply delightful view! A mother with her infant, whom she pressed delicately to her breast, and a sweet four-year-old girl."[4]

Despite some difficulties herding the reindeer through the Hamburg streets—which, of course, furthered public notice—the "Lapland" exhibit was quickly organized in the back court of the Hagenbeck property. The group of visitors simply set up their tents, corralled the reindeer, and proceeded, according to the story, to go on with their lives in this foreign land. The Hamburg public arrived in droves, and Hagenbeck quickly opened an additional entrance to the property and hired security guards to control the throngs. The idea of being able to visit a group of *people* from the far north whose manners, experiences, and entire style of life were so strikingly different from those of the city dweller of northern Europe apparently stimulated an unforeseen curiosity. The simplest—and yet to the Europeans, perhaps the most exotic—of Laplander activities seemed unimaginably fascinating to the show's visitors. With presumably an unintentional ironic contrast, Hagenbeck concluded: "A great interest was awakened every time the reindeer were milked, and a sensation very nearly developed whenever the little Laplander mother, in all her naivete and totally undisturbed by the presence of the crowd, gave her infant her breast. Our guests were unadulterated people of nature who still knew nothing of Europe's over-veneered politeness and who, deep in their souls, must have wondered what should be so fascinating about them and their simple occupations, which were by no means skills."[5] Within a few weeks, after "all of Hamburg had seen our Laplanders," Hagenbeck took the group to Berlin and Leipzig. Unfortunately, however, according to Hagenbeck, poor weather conditions and the wrong time of year resulted in the tour's barely earning enough to cover its expenses.[6] Nevertheless, the first "people show" was sufficiently successful to inspire Hagenbeck to plan almost immediately for the next tour.[7]

His first step was to send a contract to one of his animal catchers in the Sudan, Bernard Cohn, asking that along with his next transport he should "organize a number of really interesting natives . . . and bring them, along with their animals, tents, and household and hunting implements to Germany."[8] In the summer of 1876, Cohn arrived with the second show. Especially remarkable in Hagenbeck's account were two members of the troupe, a "colossal" young man nineteen years old who "caused true devastation in the hearts of European women and who himself seemed hardly immune to the attraction of the pale-faced beauties" and a young and, for Hagenbeck, quite beautiful woman, Hadjidje, who Hagenbeck believed was the first female "Nubian" seen in Europe.[9] Whereas the Sami were especially fascinating for their handling of reindeer, this show and future exhibits of peoples from the Sudan focused on the expert riding of horses and camels and the men's skill in hunting. According to Hagenbeck, "in its totality, the presentation of the caravan amounted to a sensation of the first order. Decorated only in their wild personalities, with their animals, tents, and household and hunting equipment, the guests offered a highly interesting, anthropological-zoological picture from the Sudan."[10]

Hagenbeck engaged the group of Sudanese to return to Europe the following summer, and the success of the first year's visits to Hamburg, Düsseldorf, Breslau, and Paris was repeated in the fall and winter of 1877–78 with shows in Frankfurt, Dresden, London, and Berlin, where sixty-two thousand people came to the show in a single day.[11] In the spring of 1877 Hagenbeck began to organize his fourth show, sending a young Norwegian named Johan Adrian Jacobsen, who would become the firm's most important procurer for the shows, to Greenland. With the cooperation of the Danish government, Jacobsen engaged a troupe of six for the tour—a family of four and two single men—and returned to Hamburg with the group and, Hagenbeck reported, a "highly interesting ethnographic collection." Hagenbeck noted that, in addition to the "Eskimo dogs used to pull the sleds, the households, the tents, and weapons" that he expected to come with the people, Jacobsen brought along "two kayaks, those familiar hunting boats of the Eskimo, a large woman's boat, the 'Umiak,' many interesting pieces of clothing, and many interesting implements such as snow knives, seal catching apparatus, and primitive weapons."[12]

Hagenbeck argued that such an exhibition of a people from Greenland had never before been seen in central Europe,[13] and, with its *pièce de résistance*— the men demonstrating their hunting skills in their kayaks, especially the righting of capsized boats—the exhibit found the same sort of success as the earlier exhibits. Perhaps more important, however, the show received the growing enthusiasm and support of the German academic community. Unlike the earlier shows, which had usually been presented in open fields and exhibition halls, the "Eskimo" exhibit found its home in the major German zoological gardens and received close inspection and eager backing from zoological and anthropological societies. Thus, in March 1878—thirty years after Carl Hagenbeck's father successfully exhibited six seals in a Berlin rocked by revolution to an interested, if possibly distracted, audience—the "Eskimo" troupe arrived in the now imperial capital, and the kaiser himself visited the show, admiring time and again as Ukubak, the father of the small family, righted his kayak in the water.[14]

By the end of the 1870s, it appears that Hagenbeck had weathered the more difficult years, and by 1880 the company had been given a significant boost by an exploding demand for Indian elephants spurred by the growth of large traveling circuses and the competition for ever larger numbers of elephants between Barnum and the other great circus men of the period, especially Adam Forepaugh. In 1883 alone, for example, Hagenbeck imported some sixty-seven elephants from Sri Lanka, and, along with the animals, he procured ever larger and more extravagant participants for his ethnographic exhibitions. Each "Ceylon" show, a new staple of the Hagenbeck people shows (no fewer than eight groups from Sri Lanka, for example, toured Europe under the Hagenbeck banner between 1882 and the end of the decade), seemed to outdo the one before, with one show exhibiting as many as two hundred participants. These shows often bore only a slight resemblance to the first humble "Lapland" exhibit of a decade before. Hagenbeck's description of the "Ceylon Caravan" of 1884, for example, demonstrates clearly just how far the Hagenbeck shows had evolved during their first ten years. The show consisted of sixty-seven men, women, and children and at least twenty-five elephants ranging from young calves to experienced working adults. In this exhibit, Hagenbeck noted, his intention was to catch not simply the "colorful, picturesque" of "Ceylonese" culture but also "a glimmer of its mystique."[15] According to Hagenbeck, "the colorful and fascinating scene of the camp; the majestic elephants (some decorated with shining gold saddle cloths and others equipped with working gear, carrying gigantic loads); the Indian magicians and jugglers; the devil-dancers with their grotesque masks; the beautiful, slender, doe-eyed Bajaderen with their dances which awaken the senses; and finally the great religious Perra-Harra-Parade—all of it worked a virtually captivating magic which fell upon the audience."[16] During the stay in Vienna, the exhibition was visited for three-quarters of an hour by the Austrian emperor, and on its first Sunday there, the ticket counters had to be closed twice because of the huge numbers of visitors. According to Hagenbeck, crowds of such proportion had not been reported in Vienna since the World Exhibition of 1873. In Paris in 1886, the "Ceylon Caravan" received an average of fifty thousand to sixty thousand visitors every Sunday, and around a million people visited the show during its two-and-a-half-month stay in the French capital.[17] In ten short years, Hagenbeck's "people shows" had developed from a small "Lapland" exhibit presented in the back court of the Hagenbeck property in Hamburg to huge productions touring all the major cities of Europe and patronized by hundreds of thousands of visitors.

Trying to understand the curiosity of the thousands of visitors to these ethnographic exhibits, Hagenbeck concluded that, from the very beginning, the public's attention was due largely to "the absolute novelty of not only this presentation, *but of such presentations whatsoever.*"[18] Perhaps owing to the strength of Hagenbeck's own convictions—and his consistent deployment of such statements in advertisements for his shows over the years—the idea that his "Lapland Show" was something entirely new had remarkable staying power well into the twentieth century. Only ten years after this first show, for example, and perhaps recently enough to grant the reporter some authority for knowing what was happening at the time, the writer of a review of Hagenbeck's 1884 Sinhalese show wrote: "The first experiments, as also the most important undertakings in this direction, were carried out with great success, based on our knowledge, by the well-known animal dealer Carl Hagenbeck in Hamburg, and he has also organized the large 'Sinhalese-Caravan' which is traveling through the larger cities of Germany this summer and has already left in its path Hamburg, Düsseldorf, Frankfurt am Main, and Dresden."[19] Similarly, Alexander Sokolowsky, one of the firm's scientific assistants, insisted in his official history of the firm published in 1928—more than half a century after the Sami exhibit—that, despite the proliferation of such shows around the turn of the century, "Carl Hagenbeck was without doubt the first to introduce to the civilized world the idea of the people show, and, therefore, his service in the interest of popular enlightenment should not be diminished by subsequent similar efforts."[20]

Despite Sokolowsky's assertions, however, the tradition of exhibiting non-Western peoples in Europe goes back centuries, and by the time Hagenbeck began to develop the idea, shows of people had appeared fairly regularly in the major cities of Europe for centuries. Following a tradition of explorers and conquerors going back at least to Roman times, for example, Columbus returned to Spain in 1493 with seven Arawak Indians and reported that the masses of onlookers who came out to see his procession from Seville to Barcelona "appeared to believe that he had returned with the inhabitants of another star."[21] As Urs Bitterli argued in his study of the "savage" in Western imagination, in a country where the West African and the Moor had become almost a part of daily life, the people—including the king and queen—"marveled above all else at the decorative feathers used by the Indians, the color and painting of their skins, the black smooth hair which was reminiscent of the mane of a horse, and the muscular flexibility of their bodies."[22] After this exhibit, ships from the New World regularly brought back newly discovered peoples. Hundreds, for example, were brought back from the voyages of Columbus and Vespucci alone. In 1534 Cartier brought Native Americans for the first time to France, and by the mid-sixteenth century the "Indians" had become "the main attraction of theaters and processions."[23] In the latter half of the sixteenth century, "Eskimos" began to appear more regularly in Europe, brought as captives partly to prove that the explorers had indeed discovered new lands and partly as prizes and objects of study.[24] Martin Frobisher's four unfortunate Inuit, described by George Best in 1578 as "this new prey,"[25] for example, were taken by force in two visits to Baffin Bay in 1576 and 1577.[26] Throughout the seventeenth and eighteenth centuries, the flow of exotic people continued. Champlain returned from his explorations of the St. Lawrence in 1610 with the young Huron named Sauvignon, who became the talk of Paris, and Pocahontas was the sensation of the winter of 1619 in England. In the latter decades of the eighteenth century, islanders from the South Seas made appearances, the most famous of whom was Omai. Brought back from James Cook's second circumnavigation in 1775, Omai became a focus of aristocratic society, eventually being received by King George III and painted by Sir Joshua Reynolds.[27] If Omai and Sauvignon seemed to have brushed most closely with the more elevated levels of European society, however, the presence of exotic peoples among the "freaks" in traveling fairs and markets—and by the nineteenth century their nontraveling urban forms—brought non-Western indigenous peoples into contact with the broader masses of Europe. Added to the traditional appearances of sub-Saharan and North Africans, Sami, and other Old World peoples, American Indians, Inuit, South Sea Islanders, and other groups from the New World had all made appearances in Europe before the mid-nineteenth century—well before Hagenbeck's exhibits began in the 1870s.

Hagenbeck nevertheless persisted in claiming that his exhibits of other cultures represented something entirely new because, although there were many exhibitions that claimed to show "savages" from around the world, his exhibitions of *cultures* were not shows or performances in the traditional sense. Writing about his first exhibition, for example, he insists that his "guests from the far north had no idea of presenting a show and what went along with it, and moreover, there was not any performance given." Rather, "the caravan was accommodated on the roomy plot of land behind our business on the Neuer Pferdemarkt, and therefore, was completely in the open, without artistic pieces of scenery or a backdrop. We were presented here," Hagenbeck argued, "in all honesty with a picture that, albeit in miniature, was a true copy of life in nature."[28] Essentially, the difference between his exhibits and those of others, Hagenbeck claimed, was the transparent lack of artifice characterizing his shows. As Hagenbeck's early biographer Heinrich Leutemann put it in the late 1880s, "What was of this nature previously shown was either a contemptible fairground swindle, or at least so one-sided and fake that it was not worth serious attention. And if these people shows butt up again against suspicion, because everything that is successful is subject to imitation and misuse, this itself is then the result of such imitations, which in some cases do not even eschew actual deception."[29]

Still, despite what Hagenbeck and Leutemann argued, it is sometimes difficult to distinguish between the Hagenbeck shows and the many others that either preceded or ran concurrently with them. Eight months before the arrival of Hagenbeck's first group of Sami in Hamburg, for example, another group was being exhibited in a number of the larger cities of Europe.[30] The advertisement that appeared in the conservative Berlin daily *Die Neue Preußische (Kreuz-) Zeitung* on January 15, 1875, suggests the attractions of the show. "The first Laplanders—Arctic People," the advertisement proclaims in large bold letters, "present themselves in a North-Pole theater with their reindeer, ice-dogs, and many original artifacts." According to the advertisement, unlike so many shows of exotic people which had previously toured Europe, the "authenticity" of this exhibit's members had been verified by the "Leipzig, Vienna, and Pest academies of science." A few days later, the paper carried a longer story about the visitors, noting that while the male members of the troupe "demonstrate[d] their skills in building tents, using snowshoes, throwing reindeer lassoes, singing, and dancing," the woman "[spun] her strong sewing thread from reindeer tendons."[31] The article concludes by noting that "the Laplanders had come into enough contact with [European] culture, that, through the usual passings of the hat, they display[ed] some comprehension of [Germany's] currency." Moreover, "they thoroughly [understood] the smoking of cigars and in the appreciation of alcohol they [had] been brought unfortunately apparently quite far."[32] This concern about the disappointing process of "civilizing" which the "natives" underwent during the course of their travels would grow in increasing earnestness over the years.

In advertisements for the show, its organizers, "Herr Böhle" and "Frau Willardt," explain that their "authenticated" troupe was to be seen dancing, singing, driving sleds, and putting up and pulling down tents. These are, of course, the same activities shown in the Hagenbeck exhibit, albeit, as Hagenbeck would claim, perhaps in a less organized fashion. But Leutemann, in an article he wrote for the popular illustrated magazine *Die Gartenlaube* in 1875, while recognizing the physical authenticity of this earlier group of Sami, nevertheless questioned the authenticity of its presentations.

> For a number of years several Laplanders and one or two reindeer have been shown in Germany and in Austria. For no other reason than to create more of a stir, the members of the group are clothed like Eskimos and carry along weapons which would by no means be carried by Laplanders. What is more, they conduct themselves with such affected savagery that through the whole business the ignorant must receive a completely incorrect idea of these people. Because, it seems, scientific societies, as well as individual scholars, could not doubt that they really did have Laplanders before them—even though, as it happened in Leipzig, they explicitly noted the outward forgery—out of this apparent recognition, publicity was generated and the public led into all the more confusion.[33]

What was unusual about the Hagenbeck shows, Leutemann and Hagenbeck contended, was that they were the *real thing* and that the members of the exhibits could be relied on to be only natural in their behavior. Thus, whereas Herr Böhle and Frau Willardt's "Laplanders" had been dressed in sham outfits, asked to perform inauthentic stunts, and had become all too "civilized" during their travels by taking up the smoking of cigars and the drinking of alcohol, Hagenbeck's "unadulterated people of nature" in his "Lapland" exhibit were as comfortable, Hagenbeck insists, nursing their children before the German public as they were milking their reindeer.[34]

The repeatedly emphasized claims about the naturalness or true-to-life quality of the shows point directly to the principal catalyst behind the shows in the latter half of the nineteenth century: the consistently escalating desire of Germans—and, indeed, Europeans more generally—to learn as much as possible about other peoples of the world. That desire, of course, was itself driven by a wide spectrum of forces ranging from the rarely simple curiosity that makes us all interested in people who seem somehow "different" from us to the practical needs of both colonial administrators and colonists themselves to learn all they could about the areas claimed by the various European states. It is, therefore, not much of a surprise that specifically in the second half of the nineteenth century—during a period in Germany, for example, in which the state fundamentally changed from a predominantly inward-looking confederation of largely independent small political entities to a remarkably powerful and politically unified colonial empire—the European publics wanted to know everything possible about the people within and beyond their borders.

To suggest that the Hagenbeck people shows were the product of some generic and widely expressed—and thereby somehow understandable and harmless—interest in finding out about other people does not really tell the whole story, however. Similarly, to say that Hagenbeck's shows were simply one more instance of what would become a broadly popular form of public entertainment also misses the point. Indeed, however much Hagenbeck shows resembled a great many other exhibits, it was precisely the assiduously sought after authenticity of the Hagenbeck people shows—a claim that, of course, Hagenbeck himself repeatedly made—which in fact distinguished them from other similar efforts in the period. In the end, moreover, it is that very quality of authenticity or genuineness which makes the legacy of the Hagenbeck exhibits so much more significant.[35] People understood, for example, that when they went to see the European tours of Buffalo Bill's Wild West Shows, they were not watching the *real thing* but rather a highly theatricalized version of it. Bill's shows, with their reenactments of buffalo hunts and attacks on stagecoaches by savage Indians, were so much more clearly theater than were Hagenbeck's exhibits—a point made even clearer by the weighty and ponderous messages about western progress and tragedy which were so deeply embedded in Bill's exhibits. Similarly, when people visited the "Midway Plaisance" at the 1893 World's Columbian Exposition in Chicago—with its "Beauty Show" of forty women from around the world; its palaces of Persia, India, and Algeria; its Java Village, Colorado Gold Mine, Donegal Castle, Ferris Wheel, Cabin of Sitting Bull, Eiffel Tower, and Hungarian Café; its Lapland "exhibit" and performances of Buffalo Bill's Wild West and Hagenbeck's trained wild animals—it was clear they had entered an amusement park full of the wild, the unbelievable, the curious, and the strange. In contrast, visiting the "Somali Caravan" at the zoological gardens in Frankfurt in the early summer of 1889 was like entering a foreign land. According the *Frankfurter Zeitung und Handelsblatt,* the people in the exhibit were "in fact, not performers, but rather present[ed] only that which [was] typical for them," and visitors to the exhibit should expect to see only "a picture of those manners and customs necessary to the Somali lands, and, for example, not expect that the Somali riding scenes be like those of the Bedouins and massive processions of the Sinhalese shows." The paper explained that the exhibit offered a remarkable opportunity to learn about another people and concluded:

> The quiet, careful observer who is interested in and knowledgeable about the habits of foreign peoples will find among the Somalis not only hours-long conversation, but also valuable expansion of their knowledge. We need to draw attention only to the fascinating picture of family life shown to us by the mother busy weaving at whose feet sits the dearest brown boy. To the right of the woman, the leather worker, who untiringly needles and sews, has set himself. To the right and left of these two stand or sit the other women, most of whom are also engaged in weaving. Or to point to another no less interesting picture: the smith crouches on the ground, brings the iron to a red glow, and hammers the spears straight. Or the men sit together in a circle, smoke and converse, or the food is prepared. All of these are pictures which show the visitor the Somali in their normal activities. Hundreds of visitors have already given the dances and songs of the brown company much less of their attention than the quiet activities outside of the actual presentation.[36]

So different from visiting a sideshow with its hawkers and loud and exciting adventures, visiting the Somali village—according to this report—should be imagined more as a visit to a quiet working community.

Because the Hagenbeck shows were based on the idea that what they presented was the genuine article—and here the genuine article included native peoples from northern Europe, Greenland, Labrador, North and South America, India and Sri Lanka, Mongolia, Burma, Russia, and North, West, and East Africa, among other places—interest in them crystallized almost necessarily, it seems, around the burgeoning societies for the study of anthropology, and the leaders of those societies became some of Hagenbeck's most enthusiastic supporters.[37] Indeed, as much as the needs and expectations of the leadership and members of the zoological societies for professional animal dealers who could be depended on to deliver specific, particularly valuable animals fueled the rapid growth of the Hagenbeck animal business, the needs and expectations of the leadership and members of the anthropological associations for reliable suppliers of what one observer termed "human material" stimulated the development of the Hagenbeck people shows.[38] From the enthusiasm of the anthropological societies flowed the enthusiasm of the more general populace, which had become suspicious of the edifying value of the more traditional exhibits of strange people. Of course, with the active involvement of the societies and their increased professionalization as the century wore on, the border between a faithful representation and a hoax became more and more closely guarded. Although it may have been perfectly feasible to exhibit unusual looking people as all manner of things at the beginning of the nineteenth century, by the century's close such exhibits were quickly decoded as fraudulent: only those shows that were verified as bona fide depictions received a stamp of approval from the anthropological societies and their associates at the zoological societies whose gardens became the primary sites for Hagenbeck's exhibitions of people.[39] Hagenbeck's close attention to the desires of the educated elite and its semi-professional associations elevated his exhibits well above the fairground side-show to an institution that would come to have a profound effect not only on the practice of anthropology in the decades before World War I but also on the larger cultural assumptions about "primitive people" in this crucial period. Because the wider public importance of the Hagenbeck people shows derived largely from their appeal to the academic societies, we should first direct our attention to those societies and their peculiar interests in furthering Hagenbeck's efforts.

Reporting to the Berliner Gesellschaft für Anthropologie, Ethnologie und Urgeschichte (Berlin Society for Anthropology, Ethnology, and Prehistory) in 1878, its chairman, Rudolf Virchow, brought to his colleagues' attention a particularly noteworthy appearance of a group of Greenland Inuit in the city.[40] Seeing an opportunity to carry out extensive scientific investigations, Virchow explained: "In a few words, I would like to draw the attention of the members to the Eskimos who are currently in Berlin. It is one of the most interesting ethnological scenes which unfolds before our eyes, and, I must say, one of the most strange that one can see. Over the course of the last years we have had a considerable collection of foreign specimens—in part right here in our own quarters—but none of them comes even close to the Eskimos. They present in their own way a quite strange and extraordinarily surprising appearance."[41] Within the context of the Hagenbeck exhibitions of people, however, Rudolf Virchow's legacy stretches far beyond a short introduction to a show visiting Berlin in 1878. Of course, even a cursory investigation of the shows reveals strikingly broad scholarly interest in the exhibits. For scientists of all kinds— not simply physical anthropologists and physicians but ethnologists, linguists, musicologists, and all manner of specialists, including at least one expert in indigenous shoes[42]—the exhibits provided repeated opportunities to investigate little-known peoples both without the expense and danger of traveling around the globe and with the technical resources to complete what were judged as the most thorough examinations. At a time when the study of indigenous peoples was just beginning to establish its official rules, the shows presented interested scholars with an almost unimaginably large selection of different peoples on which could be conducted the widest range of conceivable tests. But of all the scientists who dedicated a significant amount of their intellectual energy to the people exhibitions, Rudolf Virchow was their most important scientific advocate.[43] To Virchow, the shows presented such a treasure of information—whether anthropological, ethnographic, or archaeological— that over the course of some thirty years he consistently admonished his colleagues to attend the shows, lectured to the society on his findings derived from them, and defended both the exhibits and Hagenbeck in the press. Indeed, as much as anything else, it was the interest of Virchow and the Berlin Anthropological Society which granted Hagenbeck's exhibits their broad scientific authority.

In his lecture to the society about the 1878 "Eskimo Show," Virchow first notes the names and ages of the six members: Caspar Mikel Okabak (36); his wife, Juliane Maggak (24); their two children, Anne (2½) and Katrine (1¾) (fig. 22); and two single men, Hans Kokkik (41) and Heinrich Kujanje (28).[44] Despite stating that the show presented "one of the most interesting ethnological scenes which unfolds before our eyes," Virchow took note of very little about the actual exhibit. After relating briefly that "we see them in their clothes, with their dogs and equipment, their huts, their sleds, and boats, in true activity on water and on land," and repeating the old complaint that unfortunately the group's travels in Europe had "influenced in many ways their habits and ways,"

Fig. 22. Caspar Okabak, Juliane Maggak, Anne, and Katrine, 1878. *Von Tieren und Menschen: Erlebnisse und Erfahrungen* (Leipzig: Paul List, 1908). Courtesy of Hagenbeck's Tierpark.

Virchow quickly moves to the point of his lecture—to present the results of his physical examination of the four adults, which he summarized in a table, the form of which was to be echoed with only small modification for close to thirty years. It listed the names, ages, body proportions, and nine different measurements of the heads, and the resulting numbers yielded a series of calculated indices for each of the adult members of the group establishing head breadth to length, nose breadth to width, and ear height to overall height. From the perspective of more than a century, Virchow's measurements appear cold and strange, while his more impressionistic responses seem to contain so much more information. Discussing what he perceived as the extraordinary separation of the eyes of his subjects, Virchow noted, for example: "They are so widely pushed from each other, that one first has to get used to it to appreciate the otherwise in no way unpleasant physiognomic form of the face. The apparent good nature of the people can be seen clearly in their eyes. In addition, their more cultivated state contributes not a little to our appreciation of them. Their intelligence has been illuminated by the fact that they already speak a few words of German, know our money naming the individual coins, etc. They are, therefore, acclimatizing themselves by the best means and only their costumes preserve for us the complete impression of strangeness."[45] Indeed, Virchow's lecture before his colleagues reveals an essentially split assessment of the "Eskimos," and this pattern was repeated in his many similar lectures presented to the Berlin Anthropological Society. On the one hand, the lecture presented the results of his scientific study of the group in order to lead his audience through the measurements he had completed. The text directed to this part of his discourse is both highly specific and technical. Virchow noted, for example:

[Okabak] possesses a head length of 195 by a width of 144mm which gives an index of 73.7. Even with the living, therefore, one arrives at an excellent dolichocephalic measure. The case is similar with Kokkik, who possesses an index of 74.7. In contrast, I obtained with the woman 77.3 and with Kujanje 76.9, that is, mesocephalic numbers (according to the German system). Because of this, therefore, I will not place the dominant character of dolichocephalism among the Eskimos beyond discussion. The height of the ear canal is quite considerable and consequently the Ear Height Index (the ratio to the greatest length) large. In the process, it was also discovered in the case of the woman that the location of the ear hole was closer to the bregma than is otherwise to be expected. At the same time, however, it was discovered that Kujanje, whose ratios otherwise also most closely neared those of the woman, had exactly the same numbers: he and Frau Okabak have ear heights of 115mm, while the other men had ear heights of 126.5 and 127mm. The Index varied between 62.8 and 66.8.[46]

On the other hand, literally outside the more technical discussion (before and after it) and incidental to his more substantial results, however, Virchow framed his discussion with a series of brief remarks about the ethnological significance of the show. Thus, Virchow concluded his lecture noting the more interesting skills demonstrated by the troupe. "Among the skills which they present," he noted, "I would like to highlight the throwing of spears, which demonstrates one of the few examples of the use of a throwing board by a primitive people."[47] Virchow distinctly separated his observations into the related but unique fields of anthropology and ethnology—the former represented to his mind the study of humans in all their physical particulars, while within the latter he included the broader cultural creations of world societies. This pattern of separation reflected the divided and often competing interests of the Berlin Anthropological Society, and in the last thirty years of the nineteenth century especially, Hagenbeck's exhibitions of people were to play a vital role in the activities and ideas of all three of the main preoccupations of society members.

Anthropology

When the people in the shows became the subjects of the Berlin Anthropological Society, Virchow and his colleagues attempted to compile as thorough a record as possible. Body parts were measured with an astonishing desire for accuracy and little concern for comfort. Professional photographers were engaged to prepare the increasingly important photographic records that were to become the most consistently reused resources of the society.[48] Charts, assembled in Paris, Berlin, and elsewhere, against which one could scientifically establish the exact coloring of skin, eyes, and hair, were employed to assist in categorizing the various morphological characteristics of the peoples. Even plaster casts were made of consenting members of the troupes. As Sokolowsky noted, "Scholars are given the opportunity to make the most exact measurements and photographs, and study the nuances in the color of the respective natives, in order to use these notes and photos for scientific purposes."[49]

As we look back, it is often difficult to grasp the earnestness with which these sorts of investigations were conducted in the nineteenth century. Even then, of course, there were those who had an uneasy feeling about the whole business of bringing indigenous people from around the world to Europe. In an 1881 article entitled "Fuegians and European Barbarians," for example, one newspaper editor complained, "I see a large and impermissible heartlessness in dragging people to Europe, about whom our officials cannot obtain the slightest assurance whether they were aware of the meaning of their step when they trusted themselves to strange people and proceeded to the other side of the globe."[50] Concurring with the article and citing the various problems with the management of the troupe of eight people from Terra del Fuego, including recurring sicknesses that recalled the disastrous deaths of every member of the "Eskimo Show" from the previous year, another editor argued, "All of this, in any case, gives thinking members of the audience of the exhibited people material for earnest reflection, and all the more so the more frequently the word 'humanity' is led into the mouth and pen these days."[51]

These voices, however, did not go unanswered. Responding to criticism in the feuilleton section of the October 21, 1880, edition of the *Magdeburger Zeitung*, which questioned the supposed educative and scientific value of the exhibits, Virchow condemned the writer and his editors for their apparent ignorance of the scientific importance of the people shows:

> The feuilletonist enjoys saying from time to time: "it is certainly very interesting," as if that were a reproach. In this case the writer appears not to have made himself completely clear that the concept of "interest" can be very diverse. While certainly much is simply interesting for the sake of curiosity, everything else, what we investigate for the sake of knowledge, of the progressive exploration of nature and man, comes closer to us essentially only because we find it interesting. Yes, indeed, these people exhibitions are very interesting for anyone who wants to become to some extent clear about the overall position of humanity in Nature, and about the development which the human species has undergone.
>
> Whoever cannot understand that, whose preparation is so slight that he cannot understand that therein lie the most important and greatest questions which the human species can even raise, whoever believes that one should be allowed to pass over such things to more mundane matters, should be the last to write feuilletons.[52]

As Virchow made clear, despite the apparently sentimental feelings of some, as far as he was concerned the scientific stakes resting on the people shows were anything but trivial. At issue was not simply the measuring of a few heads, arms, and legs because of a crude desire to discover what people looked like. Rather, the importance of all the charts and measurements was in their use in making comparisons—most notably in the indices of related measurements, for example, head length to width—which provided a scientific basis for comparing and interrelating the races and varieties of humankind. Virchow concluded his comments about wayward writers by praising Hagenbeck's efforts. In the final analysis, he wrote, there was "a positive scientific interest of the highest order connected to these presentations." Further: "Therefore, I will not let this opportunity pass without expressing publicly our special gratitude to Herr Hagenbeck and to ask him not to let such attacks deter him but to continue in the same way as heretofore to the great benefit of anthropological science."[53] In retrospect, it is admittedly difficult to work out all the aspects of what Virchow termed a "scientific interest of the highest order." At the most basic level, the significance of the shows seems to have consisted almost entirely in gathering anthropometric data. Of course, most of this material makes for rather dry reading; Virchow described in detail the lengths and diameters of his subjects' bodies and paid little, if any, attention to the vocal or other expressions of the people or their clothes or tools. In reading his accounts, though, it is clear that for Virchow and for many others the task of gathering the body data was absolutely of elemental importance—these were the data that would be used to challenge or support some of the most hotly debated scientific issues of the period, including, of course, theories of evolution and the origins and significance of races.

The data, of course, could be turned to the defense of numerous positions. For Virchow, who by temperament was more interested in making a careful contribution to knowledge about a given people than he was in drawing generalized conclusions about the importance of his findings, the data were often used to confront almost any general theory, including evolutionary theory, for which he felt conclusive evidence was lacking. For his part, in the matter of comparative measurements of peoples, there were very few occasions in which he felt the data were conclusive one way or the other. Thus, for example, Virchow repeatedly assailed the promoters of an exhibit of the last two surviving "Aztecs" which toured Europe during most of the second half of the nineteenth century. Presenting his examinations of the two people to the society—and, indeed the larger society—on at least four occasions between 1860 and the end of the century, Virchow argued that the people being shown were in no way descendants of Aztecs and were in fact "typical microcephalics"—one of the conventional features of freak shows well into the twentieth century.[54] Similarly, in the heated debate about the racial origins of the "Nubians," a frequent focus of Hagenbeck exhibits in the 1870s and 1880s, Virchow consistently pointed out inadequacies in the data to advocates from both of the main positions at the time: those who saw the "Nubians" as simply exhibiting a more "advanced" physiognomy than "the usual Negro peoples" (that is, sub-Saharan Africans) and those who insisted that the "Nubian" peoples represented an

"East Indian" presence in Africa. The physical evidence on both sides was, according to Virchow, insufficient to reach a decisive conclusion one way or the other.[55]

In the tradition of anthropometric analysis—a tradition that continues to this day—cranial measurements took precedence, especially the skull's length and width; the height of the ear; the length and breadth of the face; the height, width, and length of the nose; and the separation of the eyes.[56] After the skull, the body itself was measured, including body height, arm span, shoulder width and height, length of upper and lower arms, length and width of hands and feet, and lengths of legs (to knee or pelvis). To be sure, the measuring of the subjects did not always proceed without a hitch. Reporting his findings on a group of "Eskimos" from Labrador whom he studied in November 1880, Virchow related an event that had already appeared in the papers.[57] Standing among the Inuit group, Virchow noted:

> You have now seen how shy the daughter [of Frau Bairngo] is; she looks like a wild animal which has been caught. Her mother has none of this frightened nature, but she is so extraordinarily mistrustful that one notices with every step she takes in a place with which she is not familiar how the new environs cause the impression of highest anxiety. It was very difficult to conduct the measurements on her, which had proceeded with the others quite simply. I began with the simplest and attempted to convince her so gradually that there was nothing bad about it; but every new act immediately aroused again her anxiety, and as soon as the body measurements began, she began to quiver and fell into the highest agitation. When I wanted to determine her arm spread and stretched her arms out horizontally—something which certainly had never happened in her life before, she suddenly became hysterical.[58]

After Bairngo's outburst—Virchow claimed after years of working in insane asylums he had never seen anything like it before—the doctor decided not to continue with her examination.

With the measurement of the exhibited peoples complete, Virchow would then turn his attention to identifying the skin color of the subjects (using the French scale prepared by the Société d'Anthropologie de Paris), eye color (again using tables and models), and hair type (often using microscopic evaluation).[59] Despite the often more subjective nature of these measurements, they were, in the view of many, some of the most important findings of all. In the case of the two groups of Australian aborigines who were examined by the society in 1883 and 1884, for example, perhaps the feature about them which intrigued the anthropologists more than anything else was the extraordinary variability of their hair qualities. According to numerous authors, the hair of the subjects, which ranged from "woolly" to "straight," proved that the Australians (importantly echoing the supposed primitive nature of the fauna of that continent more generally) were the most primitive living hominids—the *Urmenschen*—because they had within them all forms of the more highly evolved and specialized races to be found around the globe. Similarly, skin color could prove a decisive factor in the identification of a specific group's affiliation with a specific race. In the case of the "Nubians," for example, even though a large amount of the physiognomic data derived from the people suggested a stronger relation to East Indian peoples than to sub-Saharan Africans, it was finally the suggestion that the "Nubians" were often only a little darker than various Dravidian peoples which facilitated the anthropological society members' acceptance of the possibility of East Indian descent. As Virchow put it: "According to their skin, they are black like the Negroes themselves. Even if their color is somewhat lighter, at essence it is nevertheless very dark. But quite as dark are many southern Arabs and not less many Dravidian peoples of India, and if one wants to let this criterion be decisive, one would in the end be thrown back on the antiquated position of combining all blacks—including those of the eastern islands. In my opinion it has not yet been generally agreed that the Chasīa and Bedauie peoples owe their dark color to the intermixture of Negro blood; it is just as likely that they had brought it with them [from other regions]."[60]

As important to Virchow and his colleagues as the measuring of the peoples was, however, it became increasingly clear by the end of the nineteenth century that perhaps the most lasting data derived by the Berlin Anthropological Society from the people shows were the extensive photographic studies that the society commissioned from such photographers as Carl Gunther. Unlike the subjects themselves, who would be in Berlin for at most a couple of months, the anthropological photographs prepared for the society became part of the institution's permanent collection and were used extensively for decades. Because the shots were taken in the relatively controlled environment at the society, the photographers could afford to take a larger number of quality photographs without concern that uncooperative field conditions would ruin the work of months. Similarly, because extensive measurements had been completed on each individual, the traditional measuring stick used in field photos was often discarded, and the photographers could both experiment with different posing techniques and direct their attention to those aspects of their subjects which were of special interest to particular members of the society. With an unabashed enthusiasm for the prospects of photography, the anthropologist Carl Stratz, for example, insisted that "neither words, numbers, nor measures, but visual material based on personal observation and scientific insight constitutes the most important basis for our knowledge of the human races." In addition to presenting carefully selected examples of skeletons, muscles, and organs, he concluded, "The ideal of the future is a work in which the most perfect representatives of the diverse human groups—in all ages and both sexes—would be presented side by side in natural nakedness."[61]

Of course, the photographers could not always achieve the ideal of "natural nakedness," and Stratz could only lament that "good anthropological pictures of the naked body belong to the great exceptions, and among these the selected

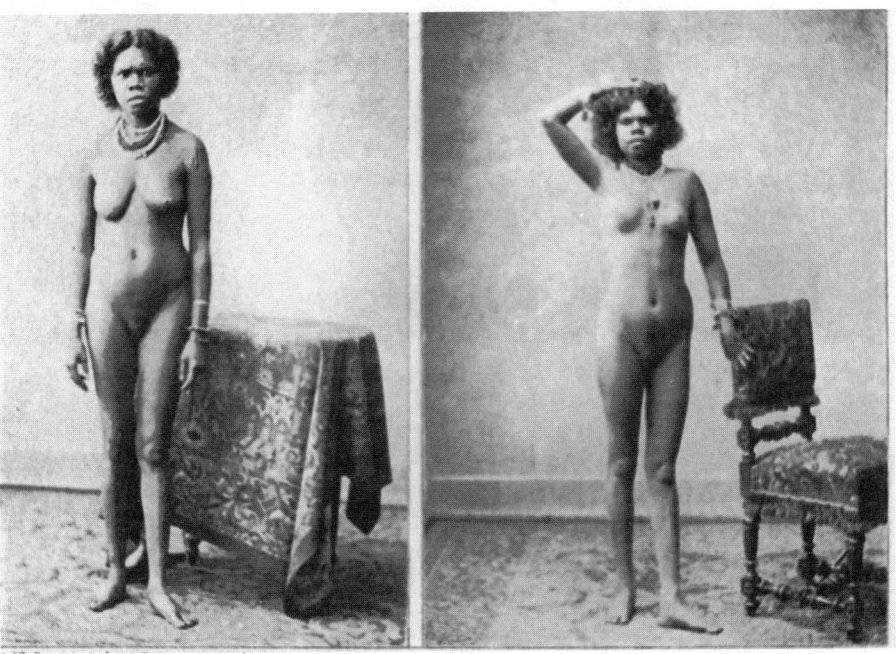

Fig. 23. The Anthropological Photograph: Australian Women, 1884. From Alexander Sokolowsky, *Menschenkunde. Eine Naturgeschichte sämtlicher Völkerrassen der Erde* (Stuttgart: Union Deutsche Verlagsgesellschaft, 1901).

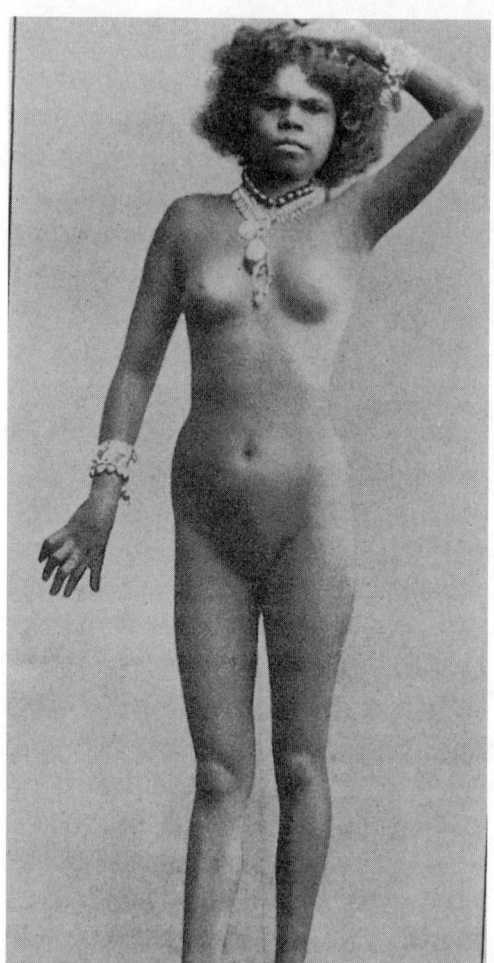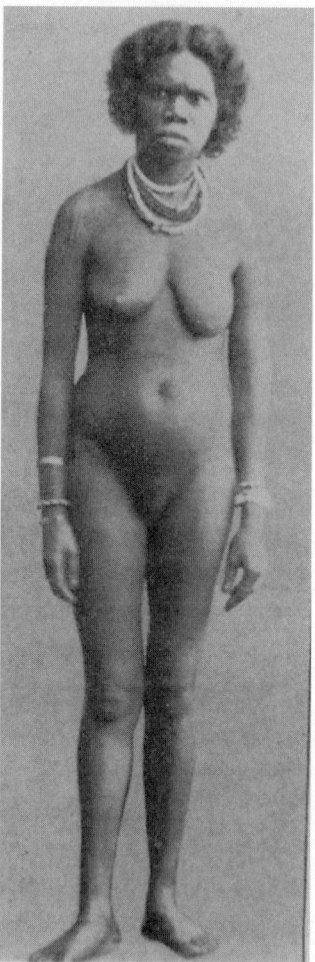

Fig. 24. The Anthropological Photograph: Australian Women, backgrounds removed, 1884. From Carl Heinrich Stratz, *Naturgeschichte des Menschen: Grundriss der somatischen Anthropologie* (Stuttgart: F. Enke, 1904).

perfect and normal individuals are even more rare."[62] Nevertheless, the people shows provided perhaps the most important opportunity that the Berlin anthropologists ever found for preparing "good anthropological pictures of the naked body." For decades, for example, Carl Gunther's pictures of the members of the shows provided a substantial amount of the visual information of anthropological texts. Such standard works as Johannes Ranke's *Der Mensch,* Friedrich Ratzel's *Völkerkunde,* and Hermann Ploss and Max Bartels's *Woman:*

An Historical, Gynaecological, and Anthropological Compendium or more popular works such as Stratz's *Naturgeschichte des Menschen* and *Die Rassenschönheit des Weibes* or Alexander Sokolowsky's *Menschenkunde, eine Naturgeschichte Sämtlicher Völkerrassen der Erde* were largely constructed around photographic material derived from the people shows, and analyses that were never completed on the actual subjects of the pictures were eventually conducted on their photographs, allowing, for example, Ploss's comparative studies of putative racial variations in the shape of female breasts and genitals.[63]

Interestingly, even apparent inadequacies in original photographs could be later corrected. In two photographs of Australian women, for example, the naked full figures were originally taken by Gunther within a studio context of Persian rugs and a draped chair (fig. 23). For Stratz's *Naturgeschichte des Menschen* (Natural history of man), however, new versions of the same pictures show the chair removed and the rugs almost completely washed out—the ideal of "natural nakedness," therefore, achieved, albeit through technological manipulation (fig. 24). The point of the pictures in Stratz's volume is the total elimination of any cultural context for the figures. These people had been removed from their native lands and photographed in a German studio, but now even the context of the studio itself has been removed. To be sure, in the anthropological works, the ideal of nakedness was necessarily reserved for representations of "primitive" peoples.[64] Thus, in a work saturated with sanitized anthropological photographs, it is important to note the typical differences between Stratz's "Fig. 327. Bavarian youth of eight head-heights" by Estinger (fig. 25), which shows a classic European nude leaning casually on a draped column, and his "Fig. 335. Blond Laplander frontal view" and "Fig. 336. Blond Laplander in profile" by Gunther (fig. 26), which show a naked man in erased space with little attention paid to ideals of either "beauty" or "modesty."

Ethnology

If the science of anthropology in nineteenth-century Germany generally meant the comparative analysis of the physical nature of humans, the study of ethnology encompassed broader "spiritual"—intellectual and cultural—achievements; in other words it examined what we generally group today under social and cultural anthropology.[65] In contrast to the anthropological examinations, the opportunities for ethnological study within the context of the people shows

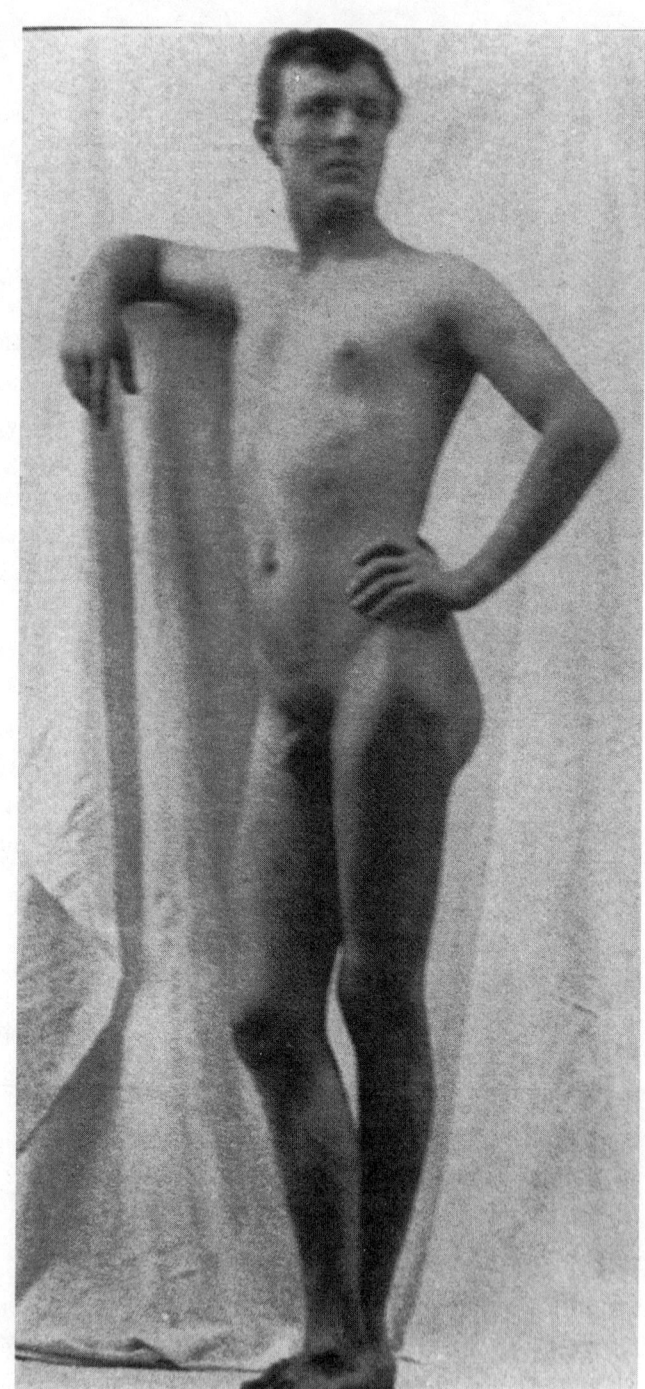

Fig. 25. The Anthro-
pological Photograph:
Bavarian Youth. From Carl
Heinrich Stratz, *Natur-*
geschichte des Menschen:
Grundriss der somatischen
Anthropologie (Stuttgart:
F. Enke, 1904).

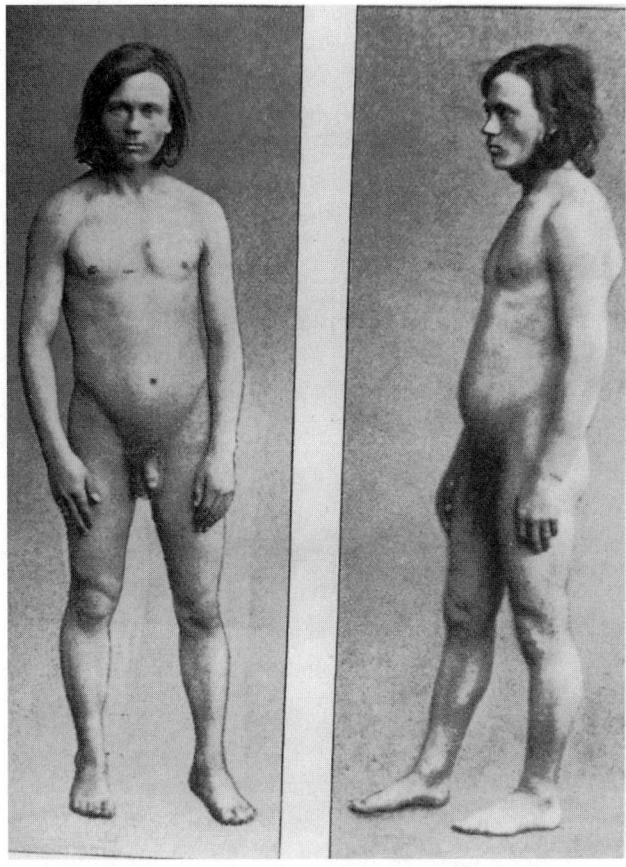

Fig. 26. The Anthropological Photograph: Blonde Laplander, ca. 1878. From Carl Heinrich Stratz, *Naturgeschichte des Menschen: Grundriss der somatischen Anthropologie* (Stuttgart: F. Enke, 1904).

were more limited for three reasons: (1) the ethnological authenticity of the early shows was often questionable and much less easily verifiable than their anthropological authenticity; (2) ethnological studies commonly required more time than anthropological examinations, and the promoters of the shows did not want their assets spending valuable exhibition time at the homes of researchers who were struggling to understand, for example, new languages; and (3) although most exhibitions included collections of artifacts that had been brought along with the "natives," these collections were often haphazardly organized articles gathered during the course of the overseas voyage from any number of places and peoples and were therefore as likely to confuse as to enlighten.

Nevertheless, it is undeniable that Hagenbeck's people shows made substantial contributions to the efforts of the ethnologists of the Berlin Anthropological Society. In terms of the artifacts alone, for example, at the exhibition one could expect to find the items that the members of the caravan brought along themselves—tents, sleds, special equipment such as saddles for camels or elephants, cooking implements, and armaments—as well as the extensive collections of artifacts gathered by Hagenbeck's agents in the lands from which the people had come. From the very first "Lapland Show," moreover, Hagenbeck made it a practice to "donate" the collection of artifacts accompanying the caravan to important museums of ethnological study; his service was recognized by his election to the anthropological societies of Berlin, Hamburg, and elsewhere.[66] Thus, all the items in the first Sami exhibit were "given"—sold—to Leipziger Museum für Völkerkunde when the tour completed its circuit.[67] Of course, all kinds of people, from sailors to colonial officials, were trying to sell artifacts to the museums during this period, but it appears that only those who had solid connections to the societies found consistent patrons. Even a figure such as Josef Menges, for example, who was a regular contributor to the scholarly geographical journal *Petermanns Mittheilungen* and who organized several shows for Hagenbeck as well as a number of others on his own account, could find it difficult to locate buyers for his collections. It seems, in fact, that Menges consistently struggled in his efforts to win the attention of the anthropological and ethnological associations and often found himself selling his collections at a loss. His exhibit at the World's Fair in Chicago in 1893, for example, was eventually sold for a six-thousand-mark loss, and in a letter of 1891 he despaired, "As far as I'm concerned, I don't want to have anything else to do with the whole business of science and am very happy that during my trips I didn't collect as much as I could have in order thereby to enrich some museum."[68] In the end, the anthropological museums and societies wanted to have a personal relationship with the collectors. In a field in which the dependability of the testimony of the collector was absolutely essential, the societies wanted to work closely with a few notable collectors, and here, again, Hagenbeck's name comes forward through his most important agent for the shows, Johan Adrian Jacobsen.[69]

As noted earlier, Jacobsen was hired by Hagenbeck to organize the first "Eskimo Show" of 1878, which was, by most accounts, perhaps the most successful show of the first five years. In the course of that exhibit's stay in Berlin, Jacobsen became acquainted with Rudolf Virchow and Germany's leading ethnologist, Adolf Bastian, and beginning in 1881, Jacobsen—in addition to working for Hagenbeck—became an official collector for the Berlin Anthropological Society. When Jacobsen returned from an expedition to collect artifacts underwritten by the Committee of Assistance for the Expansion of the Ethnological Collections of the Royal Museums in 1883, he brought with him "6,000–7,000 objects that even today," it has been argued, "count among the best that have been collected by European and American museums in Alaska and British Columbia."[70] Jacobsen maintained a close association with the Hagenbecks, for whom he worked in various capacities through the 1920s, and throughout his tenure with the company he continued to serve the interests of the Berlin Anthropological Society by organizing collecting trips to, among other places, Russia, Korea, Japan, and East India.[71]

Beyond the artifacts brought along with the shows, the members of the troupes were also occasionally engaged to inspect items already in the collections of the various museums. In one case, for example, Bastian arranged to have two of the Australian aborigines who were being exhibited at Castan's Panoptikum in the fall of 1884 interviewed about objects in the museum. The idea "was to let the Australians identify a few objects, the names and purposes of which were not yet known."[72] According to an account in the newspaper, once the doctors were convinced of the authenticity of the aborigines and that they could understand the basic vocabulary used by the visitors, the "savages"

> were led to a vitrine filled with Australian objects. With apparent pleasure the two savages looked at their native weapons—such as the boomerang and the war-club Nullah-Nullah—various decorative items, etc. They recognized a vessel made of leaves and tree bark as a honey container, called in their native language, "Kokobei." A fire starter made of two pieces of wood, which when continuously rubbed together become so hot that they are used in their home to light dry grass, appeared to arouse exceptional interest. They specified a club on the end of which was located a massive piece of crystal as an instrument which would be used only in the smashing of an enemy skull. A rod of bamboo laid before them upon which were carved diverse figures and notches they identified as a "Message Stick" used in their country for mutual communication; they [further] said that they could read the signs on one of them. The savages appeared exceedingly surprised when they were shown an Australian mummy. It was the body of an adult that had been compressed between raffia and bark to a minimum of 2½ feet in length and 1 foot in diameter and which was found hanging in a tree.[73]

Fig. 27. The Ethnographic Photograph: The Bella Coola Indians, ca. 1885. From Alexander Sokolowsky, *Menschenkunde. Eine Naturgeschichte sämtlicher Völkerrassen der Erde* (Stuttgart: Union Deutsche Verlagsgesellschaft, 1901).

Visits to the museums in the company of the "natives" tended to attract the attention of the press, but most of the ethnographic research conducted with the actual members of the troupes focused on the food, religion, work, and pleasures of the people, as well as, of course, their language. By interviewing members of the shows, linguists were given valuable opportunities to test and refine arguments about the origin and spread of the main language groups. And, of course, there were the opportunities to photograph the members of the shows going about their daily lives. Unlike the anthropological photographs, which were concerned primarily with the physical structure of the show's participants, these photographs—like the four reproduced by Alexander Sokolowsky in his study of the branches of "humanity" (fig. 27)—centered on recording such things as costumes and cultural rites.[74] But while Sokolowsky's photographs, stemming from an extensive series taken by Gunther of the Bella Coola Indians during their stay in Berlin in 1886, were used by the scientist to argue for the ethnological importance of the shows, they also clearly demonstrate the basic problems of carrying out these kinds of investigations in the context of the shows. As Wolfgang Haberland pointed out in his essay on the exhibit, the photographs of the group performing various dances and rites—including the "Chief's Dance," a healing ceremony from the "Shaman's Dance," the masked "Hamatsa Ceremony," the "Cannibal Dance," the masked "Bear Dance," the "Game for Nine Persons," the masked "Nutlomatla Dance," the "Potlatch Dance," the "Winter Dance," and the masked "Noalok Dance"—appear to focus almost entirely on the elaborate costuming, especially the extraordinary masks, which Jacobsen collected. The problem is that the costumes and masks were probably collected by Jacobsen and his brother Fillip from other tribes along the west coast of Vancouver and thus did not belong to the Bella Coola at all. "Especially noticeable," Haberland writes, "are the numerous Chilkat blankets and the Chilkat tunic worn by Ya Coutlas [the leader of the group], things hardly to be expected and never in such numbers among the Bella Coola. Most likely they were part of the collection assembled by Fillip among the Tsimshian and Tlingit." Concluding, Haberland notes that "most other parts of their costume were likewise hardly the property of the Bella Coolas . . . [and regarding] the masks and other implements, it is doubtful whether these (or at least most of them) had been brought along by the Bella Coolas."[75]

In the end, while regular attempts were made to secure substantial and dependable ethnological information from the shows by the scientific community, these efforts tended to collide with the essential nature of the shows as entertainment. Franz Boas, for example, who studied the Bella Coola in the show of 1886 earnestly, did in fact learn a great deal, but in his case as well, what he gleaned from the show was mostly his inspiration to go to the American Northwest and study the peoples and their cultures in detail in their native lands.[76] Unlike the anthropologists, who were studying a largely unmalleable object—namely, the bodies of the troupes' members—the ethnologists soon found that their studies were better undertaken in the field, where their subjects could be studied for longer periods and in their more "natural" state.

Prehistory

In a series of articles in 1878 and in his curious novel of the same year, *Rulaman: Naturgeschichtliche Erzählung aus der Zeit des Höhlenmenschen und des Höhlenbären* (Rulaman: A natural historical tale from the time of the cavemen and cave bears), David Friedrich Weinland attempted to describe life in prehistoric Germany.[77] Reading these works with their remarkable descriptions of life in another time (such as this excerpt from the popular magazine *Natur*), it is clear why *Rulaman* has remained in print ever since it was first published.

> By and large the forest was not deciduous, but rather a pine forest appropriate to the environment. Not our forest singers, the chickadee, the robin red breast, the warblers inhabited it, but only finches, buntings, woodpeckers, wild doves, ravens, jackdaws, and the like. On the mountain sides, neither the roe deer nor the proud red deer had its trails; indeed, that present was produced by a ponderous, long haired rhinoceros. A red haired elephant, the mammoth, stamped his wide path through the forest down into the nearby valley—then a swamp with many lakes, now a green meadow through which snakes a clear trout stream. On the plateau, on the other hand, were found in large gatherings the herd animals—the reindeer and a type of wild horse. In wait for them in the isolated junipers and hazel nut trees, near the few springs which animals would seek out in the evenings for a drink, lay the saber-tooth tiger, indeed more than a tiger in size and power, as we can conclude with certainty from its bones which have been preserved for us. The mighty beast would knock down its prey with a blow of its paw and—like the tiger and lion of today—lick out its victim's entrails with a barbed tongue. Sated, it would drink at the stream and then stroll with an easy quiet gait to its cave to sleep.[78]

Resonating with a broad range of popular ideas about the importance and meaning of the prehistoric German landscape and its various inhabitants and set against the vigorous controversy over evolution and what that theory implied about the origin and development of humans and human races, Weinland's stories suggest the larger popular excitement about both scientific and other investigations of German prehistory. Indeed, while Germany had become one of the most rapidly industrializing societies in the world, scientists and large segments of the general public began to imagine a past that existed far beyond romantic visions of medieval knights and even far beyond that of Tacitus: they imagined a past *before* history.

The central reasons behind this interest seem fairly straightforward. Certainly a part of the explanation is rooted in a popular antiurban, antimodern, romantic fascination with the dark woods and the dark past of Germany which was popular in the period.[79] At the same time, part of the explanation should been seen in the reorientations of science in the latter half of the nineteenth century which resulted in increasingly frequent and substantial discoveries of prehistoric remains in German-speaking countries. Although occasional extraordinary bones had always been found here and there, causing wonder, the expanding interest in cave excavations and such sensational discoveries as that of the Neanderthal Man outside Düsseldorf in 1856 dramatically drew the public's attention to human life in prehistoric times.

Perhaps the most important impetus for this new attention, however, was the 1859 publication of Darwin's *The Origin of Species* and that book's warm reception by a number of German scientists. Unlike Darwin himself, however, his most outspoken champion in Germany, Ernst Haeckel, focused his immediate attention—and indeed that of most of his colleagues—on what the theory of natural selection suggested about the development of humans from nonhumanoid species, as well as on what it implied about the development of races and cultures. As Carl Stratz noted at the turn of the century, "the various more primitive human races were examined for their resemblance to apes; . . . a list of . . . ape-like characteristics of man was compiled, and the missing link—the last connecting link between human and ape—was sought after with enthusiasm."[80] In the midst of this ardor, Haeckel made his most famous contribution to the debate about the origin of species. In 1866, he articulated what he called his "biogenetic law," in which he argued that during the process of an individual organism's development from embryo to adult, it retraced the path of its evolutionary heritage (that is, during the course of its "ontogeny" an organism "recapitulated" its "phylogeny").[81] In addition to providing a major boost to the rising science of embryology, Haeckel's law (which is with us even today in popular ideas about the presence of gills in human embryos) propelled investigations into the past of human beings because a logical corollary of the law suggested that modern Europeans—modern Germans in this context—carried within themselves the history not simply of their species but also of their race and culture. In a sense, the lives of one's ancient forebears were deeply imaginable precisely because one carried their experiences as some form of archaic inheritance.

Indeed, many German scientists in the period who studied the history of human life believed that compelling evidence was at hand to suggest the possibility of credibly reconstructing life in the prehistoric past. By studying embryology, it seemed reasonable to conclude that humans had arisen from other organisms and that, in reaching their modern form, they had passed through more primitive stages. From examining remains in caves, it was clear that long before classical times, humans had lived in northern Europe, struggling for survival with stone, wood, and bone tools and weapons. Finally, as the theory that ontogeny recapitulated phylogeny was extended, the idea that humans contained within themselves their entire phylogenetic history suggested that within modern Europeans the memories of a primitive past were present and that these memories could represent a vital part of one's nature.[82] Belief that one carried within oneself one's primitive past, however, did little to answer many questions that remained about such diverse issues as social structures, living conditions of children and the aged, forms of religious belief, development of languages, and even sexual relations in prehistoric Europe. As Weinland argued, however, although "the remains of that time are always only a few, there are . . . sufficient clues to bring back to life before our mind's eye those people of the Stone Age." Thus, the goal of *Rulaman* was to reconstruct prehistoric life in Germany. By incorporating the latest scientific research and his own carefully restrained imagination, Weinland hoped to comprehend the nature of daily life in the extreme past. He could not help but wonder, however, how people would respond to "palpable proofs of the existence and activity of a human people who lived in a time when the geological conditions of our planet were quite different—conditions, which, as we know, do not change very quickly and by which one must reckon not in thousands of years but rather hundreds of thousands of years."[83] For his part, Weinland was left with a sense of awe. Convinced that the true cave people of Europe were probably more closely related to the indigenous peoples of the European far north than to modern Germans, he nevertheless insisted that one's "spirit" and "emotions" could not fail to be aroused in learning that "here on this German mountain range, in this German valley, a Lap had once led his miserable hunter's life. And when we now visit here on the slope of the mountains, his home, his cave in which his stone and bone implements still lie, it should not be uninteresting to reconstruct a picture of that ancient, ancient time."[84]

In their efforts both to reconstruct life in the ancient past and, as Suzanne Marchand pointed out, to bridge the credibility gap yawning before a discipline lacking canonical texts and traditions, Weinland and other prehistorians sought to buttress their studies by employing the languages of more established scientific disciplines and through the argument that studies of prehistory must lead to a deeper knowledge of German—as opposed to Roman or Greek—history.[85] They met the requirement for scientific legitimacy by turning to cultures living in the nineteenth century that, they believed, seemed somehow locked—or perhaps frozen would be a better word—in prehistory. By seeking to "understand the customs and manners of living peoples of nature who can be observed in person," Weinland argued, one can "fill out the picture" and gain a deeper understanding of one's own past.[86] Sigmund Freud's opening paragraph of his 1912 *Totem and Taboo* is only one of the most famous articulations of this widely popular theoretical convention during the period:

> Prehistoric man, in the various stages of his development, is known to us through the inanimate monuments and implements which he has left behind, through the information about his art, his religion, and his attitude towards life, which has come to us either directly or by way of tradition handed down in legends, myths and fairy tales, and through the relics of his mode of thought which survive in our own manners and customs. But apart from this, in a certain sense he is still our contemporary. There are men still living who, as we believe, stand very near to primitive man, far nearer than we do, and whom we therefore regard as his direct heirs and representatives. Such is our view of those whom we describe as savages or half-savages; and their mental life must have a peculiar interest for us if we are right in seeing in it a well-preserved picture of an early stage of our own development.[87]

For Weinland and Freud, and most everyone else in scientific and even popular circles, close observation of the "mental life" of "savages" offered (among other things) an unmediated view into the life and customs of the prehistoric inhabitants of Europe.

When it came to imagining the specifically German nature of German prehistory however, Weinland and his colleagues at the end of the last century—just like the French and English—were not surprisingly selective in their choices for ancestors. The many different groups brought by Hagenbeck from Africa, for example, though repeatedly billed as "savage," were never advertised as representatives of the Stone Age. That term was reserved for peoples from colder climates.[88] Trying to meet the demand for exhibits of people who were also somehow compatible with the imaginable forms of primitivism associated with the European Stone Age, Hagenbeck, of course, usually settled on groups of Inuit and American Indians—and the more ostensibly "primitive" the better. Perhaps the most famous—the most primitive—of all of these shows, however, was the 1881 exhibit of four men, four women, and three children from Terra del Fuego. As Heinrich Leutemann recalled in 1887: "This group appeared in such complete naturalness that they had first to be given bathing suits, because with only their fur capes—their only coverings—they were not fit to be exhibited. Among the Parisians, however, precisely this abundance of naturalness met with all the more approval. The draw there was . . . massive, and it was just as large in Berlin where the Fuegians came next, for there the public flattened the barricades and it was necessary to retain security guards to keep order."[89] Trying to explain the attraction of the group, Leutemann concluded that through it "one could still see the human as he had been imagined, pushed back into the incalculable past, at the very beginning of his existence as human, after he had, that is, completely left the ape behind."[90] Photographs of the exhibit in Paris (fig. 28), as well as an illustration appearing in *Die Gartenlaube* (fig. 29), clearly demonstrate that audiences expected to see very little more at this exhibit than the Fuegians acting "savage." Unlike the huge "Ceylon Caravans" organized by Hagenbeck with their jugglers, acrobats, snake charmers, and working elephants, the "Fuegian Show" was presented without domestic animals, without an extensive ethnographic collection of artifacts, and without translators.[91] Indeed, whereas most Hagenbeck shows relied on detailing the more dramatic aspects of the cultures through the staging of elaborate ceremonies and fascinating spectacles (the dances of Sinhalese women, migrations of Laplanders, the milking of horses by Kalmucks, and camel races by Somalis), the Fuegians simply sat quietly, walked around the grounds, and prepared

Fig. 28. Fuegians at the Jardin d'Acclimatation, 1881. From Alexander Sokolowsky, *Menschenkunde. Eine Naturgeschichte sämtlicher Völkerrassen der Erde* (Stuttgart: Union Deutsche Verlagsgesellschaft, 1901).

Fig. 29. "Fuegian Types," *Die Gartenlaube*, 1881.

their food on an open fire without the use of pots. The public, however, was staggeringly enthusiastic. In Paris more than fifty thousand people visited the show on one Sunday, and at the Berlin Zoological Gardens, according to the *Neue Preußische (Kreuz-) Zeitung,* "in order to avert the earlier wild scenes of the rush of the public, a large stage some four feet in height had to be erected upon which the Fuegians were situated."[92] Most of the public was clearly more than satisfied with simply gazing upon these apparently obviously "primitive people."

The issue of primitivism also preoccupied the scientists of the Berlin Anthropological Society. In response to questions about the whether the Fuegians were what their promoters claimed, Virchow insisted in his lecture, "About the Fuegians," printed in the society's 1881 *Verhandlungen* (Transactions), that "every day" new "evidence" proved "the authenticity of the people."[93] Most impressively, he pointed to their "astonishing capacity to withstand every disadvantage of weather (despite a completely inadequate costume), an ability which has not even been closely paralleled by any other people of the Earth, with the possible exception of the Kamchadals." Despite complaints that a number of the men were coughing and two women had inflammations of their lungs, Virchow attributed these illnesses not to the extreme conditions in which the people were exhibited but rather to the fact that at the beginning of the tour in Paris "they were brought at night into a heavily heated house, but were permitted to take their usual morning baths outside. Thereby they plunged right into a pond covered with a thin layer of ice. After that they walked around almost the entire day outside most of them clothed only necessarily in small animal furs."[94]

Regarding the physical characteristics of the people, Virchow emphasized the one feature of the Fuegians which to him was most interesting: "For the first time," he writes, "I set eyes upon the foot of an adult that had never had anything on it, whether shoes, sandals, or pieces of fur. This was so because even the smallest children go barefoot even in the coldest conditions."[95] Virchow included outline drawings of the hands and feet of two of the group's members and drew the attention of his audience to a particular, and for many surprising, aspect of the Fuegians' feet: "You will notice that among these people, however low the race may be, one of the characteristics which has been argued as a mark of low evolution—namely that the second toe projects beyond the first—is not always in evidence. This phenomenon, which is noticeable among many savage peoples, is apparent among the Fuegians only in the moderation that the ends of all the toes form a slight curve, without, however, the second toe projecting noticeably forward. . . . In general, the form of the foot is rather harmonious; however, I do not want to say that it must appear to us as attractive. Our eye is so very used to the compromised foot that it takes a certain detachment to appreciate the natural proportions."[96]

Regarding the more cultural aspects of the Fuegians, Virchow had very little to say. He did note, however, that although he was not sure whether the Fuegians lived in familial groups, it was nevertheless clear that "Capitano," "Lina," and her child seemed to have established a steady relationship while on the ship to Europe.[97] In general, he concluded, "all the others interact quite freely with one another, and according to their agent, moreover, a complete communism prevails in sexual relations."[98] Admitting to the difficulty inherent in studying the social interactions of the people because of his inability to make himself understood—all efforts at communication failed, it seems, because the Fuegians simply repeated every word spoken to them—Virchow nevertheless insisted, despite both a number of his own conclusions and the general scientific discourse surrounding the people since Darwin's observations, that the Fuegians in no way represented some form of "ape-people."[99] He lectured his colleagues that, "in the case of the Fuegians, not the slightest reason exist[ed] to conclude that the race [was] by its very nature low, that they could somehow be regarded as a transitional stage between ape and man." "On the contrary," he continued, "they could have progressed further if the adversity of their environment had not repressed them so much that they remained at the lowest level of social life."[100] Using an argument rooted in an environmental determinism that was to become increasingly popular over the following decades, Virchow concluded that the Fuegians had remained in the Stone Age simply because the environment in which they lived was inhospitable to civilization.[101]

In the end, the life of the "savage" nineteenth-century Fuegians, as it was constructed by the scientific community, was completely bound to contemporary conceptions of the primitive and the relation of this primitivism to the European prehistoric past. On one level, these conceptions proved to be the antithesis of contemporary definitions of *civilization*. Intertwined within them, therefore, were expectations of loose social structures, the absence of a "feeling of shame" (clear, it was argued, in the Fuegians' lack of clothes),[102] "crude" methods of food preparation and cooking,[103] and a lack of "civilized" standards of cleanliness. Heinrich Steinitz's typical observations of the people in Terra del Fuego which were published in *Die Gartenlaube* during the show's run in Paris, for example, accentuate the oppositeness of the Fuegians when compared with civilized Europeans: "These poor wretched creatures," Steinitz writes, "were stunted in their growth; they had smeared their ugly faces with white paint, their skin was dirty and greasy, their hair tangled, their voices dissonant." Steinitz concludes (using Darwin's words) that in "viewing such men, one can hardly make oneself believe that they are fellow creatures and inhabitants of the same world."[104] At the same time, however, and on another level, this very sense of contemporary difference also paradoxically affirmed (pre)historic relation. By defining the Fuegians not as inhabitants of a different world but as "primitive" and held in that state by the suspension (or perhaps retardation) of evolutionary time, late-nineteenth-century German scientists and the general public actually recognized their connection to these people, who *were* indeed of the same world but simply from a different time—an ancient, and somehow necessarily generic, past. Indeed, the convictions expressed by Johannes Ranke in his 1887 *Der Mensch* that the Fuegians, among other "primitive" peoples, were actually startlingly similar to Stone Age Europeans were common. Ranke argued, for example:

> The Fuegians, like the Eskimos, find themselves even today in the cultural period of the Stone Age. The dog is their only domestic animal. Their tools and weapons conform to a striking degree with those recognized as coming from prehistoric Europe. As such, the Fuegians have preserved for us a deeper insight into the conditions of that period, so remote for Europe. We have learned from them in practice the way in which one made, namely in the early Stone Age of Europe, one's flintstone lances and arrowheads. The arrowheads of the Fuegians correspond to a high degree to those which our forefathers fashioned during the Stone Age. They all reveal the small, shallow, concave divots, namely on the edges and points, through which they gain a slightly saw-like appearance.[105]

The Stone Age shows were perceived—and indeed utilized by scientists—as live prehistoric displays, and as such they represent a key element in the construction of scientific and popular notions of prehistoric life in the period. At the same time, these shows themselves were understood within the context of scientific ideas about life in prehistoric Europe. Thus, in order to understand both the "primitive people exhibits" and nineteenth-century German ideas about prehistory, it is essential to understand the necessarily reflexive relationship between Hagenbeck's shows and the debates on prehistoric life: discussion about the nature of European prehistory was directly influenced by the appearance of the Hagenbeck people shows, and those shows themselves were understood within the context of developing understandings of primitive life.

How this complex reflexive relationship played itself out becomes clearer when we consider two illustrations by Heinrich Leutemann which appeared in the spring of 1878 in the German popular magazine *Natur*. The first work, "Eskimos Returning from the Polar Bear Hunt," accompanied an article written by Leutemann himself which described the group of Greenland Inuit currently being exhibited by Hagenbeck in Paris; the second, "Prehistoric Cave People Catch a Mammoth," accompanied an article by Weinland entitled "Gedanken über den Ursprung und das Leben des ureuropäischen Höhlenmenschen" (Thoughts on the origin and life of the prehistoric European cave people), which was meant to advertise Weinland's novel while providing a general introduction to the themes he presented in the longer work.[106] For "Eskimos Returning from the Polar Bear Hunt" (fig. 30), Leutemann drew an Arctic landscape and a wood lodge covered with snow.[107] In the foreground he strategically deployed a number of the props used in the show, including wood skis and a sled with a spear anchored to the back. Most of the dogs are still in their harnesses, some wearing protective boots, while one growls at the carcass of the slain polar bear. As a man and young child come out of the lodge to greet the returning group and women carry in provisions, one woman stands to the side, a baby on her back, pointing out to another child—and to us—the various activities taking place. Comparisons between this female figure, the most detailed in the work, and photographs of the members of the actual group Hagenbeck displayed make clear that the figure is based on the Inuit woman Maggak, of whom Leutemann took special note in his text. This illustration is, in fact, a combination of certain objective facts Leutemann observed about the group and his imagination of their life in Greenland. Among the items in the illustration which Leutemann actually saw, however, we should count the prominently displayed sled and spear (shown as they were photographed in the exhibit), the skis, the dogs and their harnesses, the costumes, hairstyles, and pack for carrying a small child. In short, a great deal of this illustration is simply drawn "from life."

At first glance, the illustration of the "Eskimos" seems to contrast rather markedly with Leutemann's illustration for Weinland's article on the prehis-

Fig. 30. Heinrich Leutemann, "Eskimos Returning from the Polar Bear Hunt." *Die Natur*, 1878.

toric people of Europe (fig. 31).[108] Indeed, it seems that with the exception of their creator and some shared techniques, the images of the "Eskimos" and the prehistoric killing of the mammoth have little in common. Whereas the Arctic scene shows an orderly people and a quiet event in which the only moment of tension is expressed by the curled lip of a dog glaring at the dead bear, the scene depicting the cave people in the final moments of killing a mammoth is wildly dangerous: a primordial elephant, struck by boulders, screams in agony, while the surrounding people scream in enthusiasm as one man prepares to drive a barbed spear into the head of the beast. In both scenes women and children are prominently figured, but their activities in the two works are dramatically different. In the Arctic scene women and children help with the unpacking of the hunters' sled while one mother stands aside educating her child about the bear; in the mammoth scene, however, both sexes and every age help in the hunt—indeed, in the center of the picture a young boy stands on the edge of the pit and prepares to heave a stone down on the dying primeval creature.

Weinland's interpretation of the event in his text, on the other hand, suggests more than the frenzied killing of an animal. He writes:

Fig. 31. Heinrich Leutemann, "Prehistoric European Cave People Capture a Mammoth."
Die Natur, 1878.

What an event it must have been when human cunning succeeded in catching such a forest giant! Then young and old, from small child to the old cave matriarch who walked with crutches, flooded together to see the savage scene as the poor animal, bellowing with rage and terror, was gradually tortured to death through countless thrown rocks and wounds. Indeed, compassion for animals, we must unfortunately say, has never been found either among the people of nature whom we have observed ourselves, nor among the not ignoble North American Indians. On the contrary we have found almost pleasure in killing—even in pointless murder—even as we have realized that that higher feeling [of compassion] among the children of civilized nations is hardly in-born, but rather, as a rule, appears first as a result of education.[109]

Bound together in familial clans, the cave people of Europe were imagined in a seemingly wild and yet carefully staged hunt. The mammoth has fallen into a pit, which had been covered with branches, and while the clan, under the watchful direction of the aged matriarch, distracts the creature by yelling and throwing rocks, the lead hunter with his necklace of bear claws prepares the fatal thrust of a spear precisely into the point of the mammoth's head which nineteenth-century European hunters in Africa claimed would immediately kill an elephant. While the hunt is perhaps without the "higher feelings" of compassion which Weinland ascribes to "civilized nations," it nonetheless is clearly portrayed as a cultural event rich in meaning, method, and tradition, not unlike a nineteenth-century German wild boar hunt.

Leutemann's illustrations suggest that the "living relics" themselves, as well as the basic ideas driving studies of European prehistory, could challenge dualistic paradigms of us/them and civilized/savage by positing in their stead various models of cultural evolution. In the end, only by accepting that the terms *savage, prehistoric,* and *primitive* described a wide range of cultures can we make sense of perhaps the most striking similarity of the two Leutemann illustrations: namely, that both the "Eskimos" and the prehistoric Europeans in these illustrations are shown as hunters, and specifically hunters of the most dangerous game. Although Weinland insisted (without extensive justification), for example, that "before all else, the Stone Age people [of Europe] lived from hunting," there was, in fact, precious little evidence in the European caves in which remains of humans had been found to indicate that Weinland's cave people hunted mammoths. Weinland, in fact, conceded in the text of his article that the prehistoric European preyed primarily on birds, eggs, fish, arthropods, mushrooms, fruits, and roots and secondarily on giant elk, wisent, another buffalo-like animal (the *Urstier*), and the moose and that "the great pachyderms of the German primeval forest—the mammoth and the rhinoceros—lived, by and large unharassed and peacefully beside the humans."[110] Nevertheless, in Leutemann's illustration for Weinland's article, the artist dramatically and powerfully posits the most unlikely of occurrences—the hunting of a mammoth. Far from digging up roots and searching under logs for grubs, or even chasing after a herd of reindeerlike animals, spear in hand, the European cave people—the reader was to conclude—were meat eaters who sought and killed the most dangerous of prey.

In this respect, Leutemann's illustration for the "Eskimo Show" is just as improbable as the mammoth hunt. The group shown in Paris lived before all else from the sea, and hunting expeditions to kill polar bears would have been uncommon. Nevertheless, it was precisely the most daring "Eskimo" hunting techniques that appear to have received the most attention by both the promoters (like Leutemann) and audiences of the show. Partly to get the public's attention and partly to meet the visitors' expectations that in the "fight for survival" "primitive peoples" must necessarily face down the most dangerous of beasts, shows of "peoples of nature" consistently focused on issues of hunting and eating. Similarly, while pits used to capture elephants in Africa seem to have provided the basis for Weinland's and Leutemann's descriptions of a mammoth hunt, that prehistoric Europeans specifically hunted such dangerous game seems to have developed out of a fantasy of life in the "Stone Age" which, once imposed on the both the shows and popular studies of prehistory, became remarkably self-perpetuating. Fascinated by the primitive people of the world, including their clothing, eating habits, and tools, German audiences fashioned both the "peoples of nature" exhibited in the shows and their own prehistoric countrymen and -women as wild and brave savages living in a world full of threats to their existence.

Leutemann's two illustrations are indeed projections of nineteenth-century German ideas upon both nineteenth-century Greenland "Eskimos" and prehistoric Europeans. Nevertheless, all three distinct and quite real groups are present and related to one another in each of these works. Thus, studies of "real time" similarities and contrasts (between nineteenth-century Germany and prehistoric Europe) and "relative time" similarities and contrasts (between nineteenth-century Germany and nineteenth-century Greenland) came to play significant roles in constructing knowledge not simply of prehistoric Germans and late-nineteenth-century Inuit but of late-nineteenth-century Germans as well. The actual members of the "Greenland Eskimo Show" in their interactions with their promoters and audiences, including Heinrich Leutemann, helped define the idea of an "Eskimo" in Germany; Maggak and Okabak participated in the production of themselves in German texts and minds. But the idea of an "Eskimo" remained fluid and was quickly associated with other ideas. Thus, the prominent spear in the sled of Leutemann's illustration of the "Eskimos" is echoed in the "mammoth hunt" and resonates with nineteenth-century German traditions of spear-hunting. The importance placed on ideas of family or clan, on the division of labor, on costume, and on hunting itself shows three cultures constructed by German scholars to harmonize with one another at least as much as to contrast with one another. Indeed, that the humans in Leutemann's pictures are specifically identified as such through their battles with—and, more important, defeats of—dangerous animals, in situations that were at best unsupported by demonstrated facts, makes clear that the connections between the three groups were inextricably woven and that the science of prehistory—like the sciences of anthropology and ethnology—owed more to popular ideas about the recipients of its discipline than to some detached scholarly objectivity.

That popular ideas were often deeply ingrained within the scientific investigations performed on the indigenous peoples exhibited in Hagenbeck's shows should not, of course, come as too much of a surprise. Rudolf Virchow and his colleagues were as much a part of the larger bourgeois culture of late-nineteenth-century Germany as they were part of some distinct scientific culture. At the same time, if scientists often brought popular interests to their studies, the broader public attending the shows appears to have had many very nonscientific interests in visiting the exhibits. Even if, that is, there was clearly a broader public interest in the prehistory of Europe and a certain fascination in that regard for the "Stone Age" exhibits, the extraordinary success of Hagenbeck's shows did not rest on the limited abilities of the scientific societies to impress upon the public the need for anthropometric study of the people of the world. The tens of thousands of paying visitors who arrived each day to see the people in the shows were not paying for the opportunity to conduct their scientific investigations. Even while the scientific support Hagenbeck gained from figures like Virchow was often crucial to the fortunes of his shows, it is clear that the public found its own reasons to visit the shows, and it is those reasons that would eventually lead Hagenbeck to his new ideas about exhibiting animals.

"Fabulous Animals"

In order to understand the popular side of the people shows, we must look beyond the scientific interests of physical anthropology, ethnology, and prehistory. Hagenbeck did not organize the shows to serve science; science may have validated Hagenbeck's goal of authenticity, but the shows were organized because they fascinated the more general public. It is true, of course, as one of the more recent official historians of the company has argued, that "the tours of the peoples shows . . . gave millions of Europeans their first living visual instruction about the doings of foreign peoples."[111] But what, in the end, was the nature of that instruction? In thinking about the popular reception of the people shows—and here we must consider not simply the Hagenbeck shows but also their many imitators—the most important issue to bear in mind is that despite the claims of their promoters and despite, as well, the frequently proclaimed scientific seals of approval, these shows consistently appealed to the most basic expectations of the general audience.

Indeed, a significant aspect of the verisimilitude sought in the Hagenbeck shows revolved precisely around meeting the general public's expectations for an exhibit. Thus, it is very difficult, in fact, to find a show that generally or even partially challenged stereotypes of the behavior anticipated of the show's participants: American "Indians" were required to ride horses and sometimes even scalp settlers, "Eskimos" were required to paddle kayaks, "Bedouins" were required to ride camels, the "Amazon Warriors" of Dahomey were required to be warlike. These expectations, more than anything else, dictated the narrative content of the shows. In some cases, moreover, crude stereotypes meant to degrade or at least mock the shows' participants seem to have been at the very center of the programs. A large part of the attraction for the exhibit "Prince Dido of Didotown," consisting of Prince Dido, two of his wives, and "a few other Blacks from Cameroon,"[112] which arrived in Germany in 1886, for example, was the acquisition of Cameroon as a protectorate in the winter of 1884–85.[113] Noting the instant interest in the new colony, Carl Hagenbeck's brother John recalled: "The previously practically unknown Cameroon was suddenly extraordinarily popular. Not just in the newspaper columns was it continually discussed, but also on the stage and in the magazines. Our humorists had quickly found out, what a wealth of unintentional comic material was attached to our fellow black countrymen, especially with the so-called Kings Bell and Akwa and their many associated princes. One could hardly visit a single burlesque theater, therefore, without having to put up with a Cameroon couplet; there were even pen-ready librettists and composers processing the good material as quickly as possible into operettas."[114] John Hagenbeck concluded with a keen sense of the Hagenbeck operation: "With the vivid sense for what is current which always controlled the undertakings of the Hagenbeck firm, we naturally did not let the favorable economic opportunity go unused. We put together a group of people from Cameroon."[115] Not surprisingly in the context of Germany's colonial activities at the time, this "Cameroon Show" fortified popular ideas of Africans as the press interpreted Prince Dido's understandable desire to wear a top hat, frock, and pants as typical of the essentially imitative nature of "Negroes."[116] Similarly, while Crown Prince Friedrich Wilhelm spurred everyone to laugh by referring to Prince Dido as "his colleague,"[117] Leutemann noted that "the most interesting aspect of the Cameroon Show was the presentation of the often cited and in fact remarkable drum language, through which, so it appears, all and everything can be expressed." "An amusing proof of this," Leutemann concluded, "is the drumming which has already been heard in the night in Cameroon, 'King Bell is a scoundrel, scoundrel, scoundrel.'"[118]

At the same time that the desires of the audience were rarely in jeopardy of being seriously confronted, moreover, most of the people shows by the end of the century—even of the Hagenbeck variety—began to incorporate increasingly complex narrative structures. Rather than just strolling around a village, therefore, visitors to the shows were more and more frequently presented with carefully constructed reenactments meant to demonstrate the varied nature of human experience in the exhibited region. This trend can at least partly be attributed to the desire of the "impresarios" to present as full an introduction to a region and culture as possible, but above all it appears to be the result of Hagenbeck's and his agents' desire to get their money's worth.[119] Thus, by the end of the 1870s, the performers in the Hagenbeck people shows were normally contractually obligated to present their dances, crafts, songs, and the like, and the more elaborate the presentations, the more likely the show would make a profit. Part of the Articles of Agreement of July 25, 1885, between Johan Adrian Jacobsen, nine members of the Bella Coola, and Israel Wood Powell (Indian superintendent, British Columbia), for example, reads: "Between the hours of eight and twelve o'clock in the forenoon and between half past one and six in the afternoon inclusive of Sundays according to the best of his skill and ability in concert with the other parties of the first part and under the direction of the said John Adrian Jacobsen exhibit himself before the Public in the performance of Indian games and recreations in the use of bows and arrows in singing and dancing and speaking and otherwise in showing the habits manners and customs of the Indians and will during the term of the engagement hereinafter mentioned behave himself reasonably and respectably."[120] In the context of these performances, scientific accuracy necessarily played a secondary role; indeed, the rather limited financial success of the "Bella Coola Show"—which was widely regarded for its ethnological significance—suggested that the audiences had little taste for watching enactments of difficult to understand, if also ethnologically interesting, rituals.[121]

The "Indian Caravan" organized by Carl Hagenbeck's half brothers John and Gustav is typical of how the project of re-creating a convincing realism evident from the very early Hagenbeck exhibits could be adapted to the increasingly performative nature of the exhibits. The first paragraphs of a small pamphlet written to introduce the show described what visitors would likely see:

> In this exhibited caravan consisting of about 50 people—men, women, and children—stand out especially: the tricksters, acrobats, and jugglers of the Maharadis and Bandscharis peoples coming from the state of Travancare; Tamils from the coast of Malabar; Indian Temple Dancers; Bajaderen and musicians from Tanjore; Muslim Indians as fakirs, snake charmers, and magicians; Devil-dancers and Fire-dancers; and even a Muslim dwarf from Madura.
>
> Constructed of authentic Indian huts, the Indian village which unfolds before the eyes of the audience offers a charming sight. At the same time the public is offered an opportunity here to purchase genuine Indian crafts such as boxes, blankets, bowls, as well as curiosities, shells, lace, etc.
>
> The temple which rises from the middle of the arena, an accurate reconstruction of a Brahman Gopura in Madura (Southern India), which is adorned with pictures of the Gods taken from the teachings of Medas and many other decorations, gives evidence to the extravagant fantasy of the Brahman cult.
>
> The Brothers J. and G. Hagenbeck have endeavored first of all to bring to life an interesting picture for the concerned public, which presents the life and ways of the

Malabar tribes, and have further given themselves the task to show something absolutely worth seeing—as has as yet never been offered by any other group—to the interested visitors.[122]

The pamphlet emphasizes the real locations from which the people in the show were alleged to have come; it highlights authentic huts, genuine crafts, and accurate reconstructions, but the focus of the show—and of the pamphlet—was the exoticism of human life in India. Consequently, the picture painted by the show was not daily life in India but a world of musicians, jugglers, snake charmers, dancers, and a dwarf. In the end, the reconstruction of the Brahman temple decorated with images of gods was meant to catch the paying public's imagination and to show something "absolutely worth seeing." The cover of the pamphlet, executed by the press of Adolf Friedlander, echoes the theme of a view into an exotic world (fig. 32). Dressed in a red and white shirt, with a necklace and a cobra around his neck, a barefoot Indian has unfurled a poster showing life in India. In a setting of palm trees and a deep canyon, elephants heave tree trunks and snakes are charmed, dancers go through their rhythms, a zebu pulls a cart, and food is prepared in the foreground. This, then, was the world of India as conceived in the Hagenbeck people shows.[123]

In addition to the re-created villages, most Hagenbeck shows presented scheduled performances. The "Ethiopian Show" of 1909, for example, included a play with five scenes, which, while attempting to detail the daily life of a group of Habr Awal, also tapped into the romance of Hagenbeck's famous animal trade.[124] In the first scene the stage is set as an oasis where women work before their huts. In the second, a family of Habr Awal arrive, led by their chief riding a white horse, while a herd of goats and dromedaries packed with assorted materials for huts and households are led into the village. The third scene begins with the arrival of a Hagenbeck animal caravan on its way to the coast. Camels appear again carrying cages for animals, and after being greeted by the Habr Awal chief, Hagenbeck's agents try to purchase a captured zebra. Soon sentries are posted, and all settle down for a rest. The action builds in the fourth scene when the camp is attacked by another tribe. The guard is killed, and the animals are stolen; after a violent battle, the bandits retreat, leaving their dead and wounded. According to the pamphlet detailing the show, "The wounded are slaughtered by cutting their throats." Victory dances begin among the Habr Awal, "wild shouts of Uh-hu-hu are heard, the cutting of throats is

Fig. 32. Adolf Friedlander, illustration for Indian caravan show, 1884–85.

threatened, and spears are thrown up and caught again." The fifth and final scene shows the "Palaver" and eventual peace between the Habr Awal and the Isas and Danakil peoples. According to the pamphlet, "The stolen animals are returned. . . . The Habr Awal pay blood money (two dromedaries and a girl)," and the chief of the Habr Awal makes a final address.[125] It should be clear that the drama presented as part of the "Ethiopian Show" had little to do with the anthropology of the region, and the ethnological significance of the event was limited at best. Instead, the show seen by the German public at Carl Hagenbeck's Animal Park in Stellingen focused on the romantic, dangerous—and, of course, largely fantastical—lives of Hagenbeck's animal dealers.

Claiming to offer real experiences with exotic people while generally reinforcing stereotypes of these same people, the shows were often, therefore, more about the nature of European civilization than they were about the exhibited peoples of the world. The critique of civilization worked in two directions, however: civilization could either be claimed as the triumph of progress or be found wanting beside the "naturalness" of the peoples exhibited in the shows. In either case, though, the participants were clearly actively "created" out of the viewers' own obsessions and fantasies. That a viewer might despair the value of "civilization" and praise the "innocence" of the members of an exhibit does not, of course, mean that the resulting critique would manage to avoid the larger cultural presuppositions about the exhibited people. For example, although Peter Altenberg's short but remarkable book *Ashantee,* which was based on an 1897 non-Hagenbeck "Ashanti Show" at the zoological gardens in Vienna, condemns European culture and celebrates the "purity" of the exhibit's members, it remains all the while thoroughly suffused with the racism it attacks.[126] During the course of the visit of the show in Vienna, Altenberg, who was locally famous for enjoying the less wholesome aspects of urban life, became intimately involved with various female members of the troupe. These involvements served as the backdrop to his examination of hypocrisy of his fellow Viennese. In a contemporary review of the work, Max Messer stressed that in *Ashantee* "the *problematic of culture* has been shown from personal experiences and before all else, the lie has been put to rest that culture and humanity are identical concepts. Peter Altenberg teaches us to see culture through the lens of 'unculture.' While we look at ourselves through the eyes of those black, simple, childlike humans, we see ourselves more clearly, catch sight of previously obscured inadequacies in our lives; grasp our great objectives more meaningfully, and shorten and ease the laborious paths."[127]

From our perspective, it is difficult not to see that Altenberg, who supported himself in a sporadic way from his writing, characterized the participants in the "Ashanti Show" through an infantilizing perspective deeply rooted in the racism he sought to critique. Nevertheless, his observations on the interactions of the Viennese with the members of the exhibit provide a great deal of insight into the public's response to the exhibit. In a small piece provocatively entitled "Culture," for example, Altenberg describes an evening when he and two members of the show, Big Akolé (aged 17) and Bibi Akolé (7), were invited to dinner in the city. After a dinner at which the "natives" demonstrated their polished etiquette, each was given a French doll:

First they sang them to sleep and kissed them.

Suddenly the big Akolé let her toga fall from her ideal upper body and gave her breast to her doll. Little Akolé stood there deeply despairing with her hungry doll in her arms.

Frau H. told her guests that it was the most sacred sight of her life.

The guests all agreed, if not so bombastically.

Even Monsieur R. de B. smiled in a way that one does not really smile when one smiles.[128]

Altenberg's vignettes point to the gulf of misunderstanding at the heart of the shows which the promoters did little, if anything, to bridge. Indeed, they point to the inherent contradictions in the whole project of re-creating "primitive worlds" in the heart of European cities. Just how lacking in subtle observation the organizers of and visitors to the shows could be is evident throughout the reports of the exhibits. The satiric and condescending tone of a report in the *Frankfurter Zeitung und Handelsblatt* of the arrival of a Kalmuck show, for example, seems in many ways typical: "In our Zoological-Ethnographic Garden the Kalmucks have followed in the footsteps of the Samoans. It is only lucky that the Mongolian people of the steppes did not follow on the Samoans' heels—the slit-eyed Kascha wear such massive boots and the beautiful Fay have such delicate bones. From the standpoint of comparative Sunday-afternoon-anthropology, that says about everything: so graceful as is the whole manner of the brown children of the tropical island, so less elegant is the nature of the yellow nomads whose so-called cradle lies (or better said, is ridden through) at the mouth of the Volga. Indeed the Kalmucks are a riding people through and through." After detailing the limited possessions of the Kalmucks—their horses; "riding gear . . . ; several camels and sheep; a few tents and coverings; the rarely changed clothes and ditto frugal underwear for each member, whether man, woman, aged, or baby; a tobacco pipe; a half dozen whips; and some odds and ends for religious purposes"—the reporter returned to his distaste for the people. He writes: "At all costs we do not want to get too close to these Kalmuck people, certainly not, not for everything in the world. . . . In any case, it is not out of the question that such a show can nevertheless be entertaining and informative. The program of the presentation is extensive enough. . . . The lassos are skillfully thrown by the men. Almost every time the noose falls around the neck of the camel or horse which is to be captured. The proficiency in the beautiful arts, in song and dance, however, is understandably not very developed. Arid and sleepy as the endless barren steppes, the melodies of the song or the plucked fiddle drag on and on and the dancing pair develop little more fire than a burned out match in a pond."[129] It seems, in fact, that *mis*understanding was an integral part of the success of the people shows. Reporting on an occurrence at the exhibit of "Australian Cannibals" in Castan's Panoptikum in Berlin, the "about town" section of the *Neue Preußische (Kreuz-) Zeitung* enthusiastically reported, for example:

> An uncouth woman from Berlin could not deny herself the chance to grab quite firmly the arm of one of the brown fellows who strolled among the public. Caught by surprise, he whirled around, and, for fun, the woman hit him softly on the arm. The cannibal, however, did not get the joke and before she knew it, she had received a rough slap on her hand from the muscled hand of the boomerang thrower. The woman appeared to believe that this too was meant in fun; in any case she returned the hit, certainly with no intention of measuring herself against the warrior in a fight. That was too much for the cannibal. His in any case not classically beautiful face screwed up into an infuriated grimace, and he had already raised his arm for battle, when fortunately the master of the savage horde came up and brought peace. The idea of smoking a peace pipe was rejected.[130]

The idea that one of the Australian "cannibals" almost attacked a "civilized" white woman who could not restrain her desire to feel the skin of the "savage" could only contribute to the Berlin public's general enthusiasm for the show.

The woman's desire to touch the "savage"—and the newspaper's enthusiasm to report the event—point, of course, to what is undeniably one of the most important forces behind the success of a great variety of Hagenbeck's exhibits: from the very first shows of the "Laplanders" and the "Nubians" to the very last of the South Sea Islanders, the shows provided their European audiences with the opportunity to look at, and occasionally touch, the often only slightly clothed bodies of the exhibits' members. A newspaper article describing a performance of Hagenbeck's "Amazon Corps" show of 1890 in Umlauff's "World Museum" makes more explicit the often highly eroticized nature of some of these shows.[131] In a room filled with stuffed animals from Africa and where walls were covered with scenes from the jungle and the veldt, the public stood closely packed together before the decorated stage "on which the Amazon corps, the body guard of the Kings of Dahomey—present[ed] their marches, dances, and arts of war in attack, battle, and thunderous victory celebrations. A dozen brown-skinned beauties in fantastic costumes of shells and corals—otherwise, however, practically naked—demonstrate[d] in the company of magnificent, muscled warriors what at home [brought] the King fame and glory."[132] After a performance filled with swords and shields crashing together and "blood-thirsty" screams, the "Amazons" sat down at the front of the stage. The audience pushed in close, and the women in the show "smile[d] at everyone who [spoke] to them or through sensual curiosity [let] his hand glide over their beautiful, tender skin." Soon the leader of the "Amazons," Gumma, entered the crowd with her comrades and began to offer souvenir pictures of the group to the enthusiastic audience. The article continues:

> "How much does the picture cost, beautiful warrior?" Gumma understands only "cost" and smiles: "One mark!" "Isn't that a little expensive, Gumma?" She shakes her head and the shells rattle softly. For a moment longer she holds the picture out before the eyes of the interested party, then she quickly turns her gleaming brown shoulders and walks slowly on. If the interested person is a man, however, she still holds a magic cure ready. The usual pictures show her in a circle of her warriors, the breasts chastisingly covered with a hanging drape of shells and beads. She possesses in addition, however, single pictures of herself in which the drape is missing. She now pulls such a picture out in front of the others from under the stack. "Two marks!" she says through her nose while her beautiful countenance betrays not the slightest movement.
>
> And whereas before one mark appeared exorbitant, this doubly large demand is simply a giveaway price. Gumma understands the white men of creation already thoroughly and treats them accordingly.[133]

It should not be a surprise to discover that part of the enthusiasm for the people shows stemmed from a veiled and sometimes blatant erotic voyeurism.[134] Simply, the shows provided a purportedly legitimate arena for men and women to see and to "study" sometimes almost completely naked men and women, and from the very first two shows, which "featured" the exposed breast of the nursing Sami woman, the beautiful Hadjidje, and the "devastating" nineteen-year-old Hamran warrior, the confluence of nakedness, beauty, and exoticism was to play an important role in many of the shows. In some cases the reports of the shows talk of the attractiveness of people in modest terms. An editor with *Frankfurter Zeitung und Handelsblatt,* for example, noted the arrival of a group of Samoan women in 1896:

> When the announcements speak of "Samoan beauties," and when Rudolf Virchow with the weight of his name has given evidence of the extreme grace and deftness of

the people, not a single word is exaggerated. The brown beauties in light, colorful cotton costumes with broad red, blue, or green silk sashes and pleasing, groomed, black hair decorated with flowers are absolutely sweet characters of attractive size, gentle disposition, and lively eyes. A few of them are very nearly captivatingly beautiful. The dances which they present, the songs with which they accompany the dance or the preparation of the Kawa, a root preparation which they drink not without offering their audience a toast—every detail of the exhibition appears not only unusual but also appealing to our sense of beauty.[135]

In a sense, though, even this relatively innocuous report makes clear that the point of this exhibit was the opportunity to stare at beautiful and exotic-looking women. Although, for example, there were men in this show, the article focuses on the "beauties," on their clothes, their hair, their eyes, their dances. That Samoa had become a German protectorate further increased the potential appeal of this exhibit, which emphasized the possibility of paradisiacal scenes in the South Seas. Although the language of this account is relatively discrete, however, other accounts suggest the more explicitly sexual appeal of some of these exhibits. Peter Altenberg, for example, described being asked how one could arrange a meeting with one of the girls in the "Ashanti Show":

"Excuse me, but what about these black young girls?!— — —Yes?!"
"No."
"Oh. You're a gentleman; you won't tell anything."
"I don't have anything to tell"
"O.K., then, do they take money?!"
"Yes."
"And silk scarves?!"
"Yes."
"And then what?!"
"Then nothing."
. . .
"But I heard that one could buy young black girls?!"[136]

Even here, of course, Altenberg is relatively circumspect in his language, even if he apparently was not with his own behavior.

Beyond the written accounts of the erotic content of the shows, there remain photographic collections stemming from the exhibits which also clearly indicate that the success of a number of shows was, at least partly, the result of sexual curiosity. As Thomas Theye noted, "for Europeans of that time, photographs of naked savages were a legal opportunity to observe the naked body," and consequently, "the studios offered an enormous number of photographs which ranged from harmless friendly poses to others which stressed a clearer sexual message."[137] In the case of Hermann Ploss and Max Bartels's popular 1913 *Woman: An Historical, Gynaecological, and Anthropological Compendium,* for example, with its hundreds of photographs—a great many stemming from the people shows—it seems clear, as Jan Lederbogen noted, that "under the pretext of science and popular education, it was possible to publish photographs of unclothed representatives of foreign peoples."[138] The wide-reaching audience for these pictures, which, of course, over the course of the past century has far exceeded the numbers of those who actually saw the shows, suggests the lasting importance of such pictures in reinforcing connections between "exotic people" and sex.[139] Of course, it was only in the interest of Hagenbeck and his representatives, as well as the promoters of similar shows, to encourage the sexual curiosity of the public. In fact, even the most outwardly innocent shows could easily profit from this kind of curiosity. For example, given the long history of imagining peoples from the South Seas in particularly sexual terms—a history that stretches back to the eighteenth century and accounts of the circumnavigations of Bougainville and Cook—we should not have to think too hard to realize that when people went to see Hagenbeck's "Samoan" exhibit, they had in mind certain particularly sexual preconceptions of the show's participants.

In his official history of the Hagenbecks from 1929, Ludwig Zukowsky drew attention to the scientific importance of the people shows: "The viewer receives at first a look into the life, customs, and ways of other peoples of culture or peoples of nature and therein lies a scientific value which raises the presentation above the level of bare curiosity. Further, however, through the shows science is delivered valuable research material. The linguist can pursue interesting language studies on the natives coming from diverse regions; the phoneticist can inform himself on living research material about the forms of articulation. For these reasons, then, scholars are standing guests at Stellingen during the period of the show. In addition the anthropologist and the ethnographer find a great deal to learn about the natives, their clothes, weapons, and implements."[140]

Another of the "house historians" from the period, Alexander Sokolowsky, pointed additionally to the positive effect that the shows had on the public's interest in foreign lands, and more specifically in the colonial movement. Sokolowsky writes: "It must not go unrecognized that the acquaintance of the masses with representatives of foreign peoples must function in a stimulating and enlightening way for colonial efforts. At the beginning of our colonial acquisitions a great many people in our fatherland had as good as no, or only very little, knowledge of the nature and human life of our protectorates. Through the presentation of the groups of peoples with their households and domestic animals, interest in our colonies was raised and understanding for colonial tasks was broadened. Not least, for many people, contact with foreign natives broke the spell which had made some people easily deceived and, through ignorance and prejudice, adversaries of colonial issues."[141] To be fair, Sokolowsky was trying to secure fame for Hagenbeck's achievements, and to a certain degree his argument should be understood as his effort to claim additional importance for the shows. Nevertheless, his general statement that the shows helped reduce the European public's anxieties about possessing colonial lands can be taken at face value. To be completely clear, there was never a show that sought to denounce the activities of Europeans abroad. Through the exhibition of the indigenous peoples of the world in the highly controlled settings of the shows, with those shows' important claims of authenticity, the larger public in the major cities of Germany and other European nations received firsthand experience with the "natives" of their colonial possessions. Thus, the shows contributed to the idea that the efforts of the colonial societies were advantageous both to the indigenous peoples whose lands were being occupied and to the Europeans who were occupying them. Not surprisingly, most observers of these shows saw happy and contented "primitives" going about their simplified lives and presenting little threat to European authority. This is precisely what one should recognize in a photograph of a "humorous moment" during the kaiser's visit to Hagenbeck's Animal Park in 1913 (fig. 33).

Clearly what visitors came away thinking after visiting one of Hagenbeck's people shows depended in large measure on which show they saw; visiting one of the Somali shows was a very different experience than visiting the Labrador Inuit exhibit. On the other hand, at the base of all the shows was an assumption that, as objects of exhibition, the subjects of the exhibits were at the disposal of those who wanted, in whatever way possible, to study them. In the case of anthropologists and anatomists, the scientists understood the shows as

Fig. 33. Kaiser Wilhelm II and Heinrich Hagenbeck, at Stellingen, 1913. Courtesy, Schulz Collection, Lampasas, Texas.

vital opportunities to carry out scientific study, and little attention was paid to reflecting on whether certain forms of study were ethically acceptable. In the case of the women from Terra del Fuego who had made it quite clear, for example, that they did not wish to have pelvic examinations, scientists such as Theodor von Bischoff felt that they had a right to carry out such examinations, and Bischoff, in the end, was allowed to carry through with his study, albeit on the corpses of those female members of the troupe who had died during their tour through Europe.[142] In general, the scientists who studied the members of the people shows placed very few limits on their investigations and systematically directed their attention to their subjects' most private concerns; focusing on the physical nature of the races, they prepared comparative studies of genitalia; focusing on their spiritual beliefs, they investigated and tried to decode the psychological and religious concepts of the peoples. At the same time, the data derived from the shows were quickly deployed in "scientific" theories of racism, including, for example, the elaborate trees of descent showing the progression of human races from the most primitive to the most advanced.[143] The right to observe, indeed to stare, which was systematically reinforced by the promoters of the people shows, was not, of course, limited to the scholarly community. The "civilized" society, scorned by Altenberg in his *Ashantee,* poked, prodded, harassed, and seduced the members of the people shows; they had been brought to Europe, after all, to entertain the public. Similarly, voyeurism—whether legitimated in the context of "scientific study" or blatantly displayed in naked and erotic photos of the members of the groups—was in the interest of the promoters of the shows and was certainly part of many of the shows' extraordinary successes.

In his memoirs, Carl Hagenbeck reflected for a moment on the fate of the people in his shows. Wondering what their lives were like after they were returned, he wrote, "You have all returned home long ago to the lands of your forefathers, and the voyage in the land of the white man—which sent you home rich with treasures—has become the great and unforgettable adventure of your life.[144] Hagenbeck's reasoning that the shows represented for the exhibited people perhaps the most extraordinary event in their lives, an adventure to be told for generations, has had remarkable staying power over the years.[145] It is, of course, undeniable that for probably the majority of those people who were brought to Europe to be shown in either a Hagenbeck or other people show, the trip brought wealth and prestige at home. Günter Niemeyer cited the return of the 1878 Greenland Eskimo group as an example of how popular the shows were with the people who were displayed: "Richly laden with gifts of every kind and with wonderful pay, the Eskimos returned to their home. Ukubak invited all his fellow tribesmen in the Disko harbor to a huge party. He had purchased so much coffee, tea, sugar, flour and zwieback that he could have hosted for days. Atop a large stone he lectured on his trip to Europe, which had, in fact, even led him to Paris; the lectures had no end. Who can wonder that no one wanted to go hunting anymore and that they all lay on their lazy skins and dreamed of travels in the South. Things got so bad that the Danish government prohibited taking Greenland Eskimos to Europe again."[146] However amusing Niemeyer's explanation for the reasons behind the Danish government's ban on taking Greenland natives to Europe was intended to be, it remains singularly lacking in reflection and taste, especially when considered in the context of the second Hagenbeck "Eskimo Show," again organized by Jacobsen, featuring a group of eight Inuit from Labrador, all of whom died of small pox during their European circuit. In his 1910 chronicle of Labrador and the Moravian missions there, W. G. Gosling condemned the practice of taking

"Eskimos" to be exhibited in Europe and America and called for legislation to prevent the practice. Drawing attention to the Hagenbeck show, he wrote that "Hagenbeck, the well-known wild animal exhibitor of Hamburg, sent to Labrador and induced eight Eskimo men, women, and children to go to Europe for exhibition purposes. The Brethren at once saw the probable evil consequences, and used all their persuasive powers to prevent them from going. But the attraction of good pay, easily earned, outweighed the warnings of the Missionaries. Their forebodings were only too quickly realized." Gosling concluded that the "poor creatures" were "butchered to make a Roman holiday."[147]

Thus, even though it is certain that many members of the exhibitions returned to their native homes with a wealth of experiences and, by the standards of their countrymen and -women, with a wealth of money and gifts, it must be stated very clearly that the shows were never organized to benefit indigenous peoples, and what wealth they gained from them was an insignificant portion of the often extraordinary profits that the shows could reap.[148] Moreover, the members of the exhibits rarely took home those benefits usually associated with "civilization." Because of the demand for authenticity—the need that the "natives" be sufficiently "native"—any form of sustained education was usually out of the question.[149] Thus, with only a few exceptions, those peoples brought to Europe to be put on display returned without substantial familiarity with European languages, the skills to read and write, or the much-desired instruction in the use of technologies such as sewing machines. Similarly, there seems to have been little effort on behalf of Hagenbeck and his agents to follow up on their charges after they had been returned to their homes to see if the tour had had any lasting negative effects on the people—for example, the transmission of diseases. The indigenous people were only valuable insofar as they remained "native," and after the shows, they were returned to their homes with little fanfare.

In his 1885 biography of Hagenbeck, Heinrich Leutemann noted: "It seems that people can always find something to fuss about and there have been voices who have wanted to describe this 'exhibition of people' as something degrading. Then one must also damn at the same time the whole theatrical enterprise and all public showing of oneself, because the difference—that in the people shows the people behaved only naturally while actors, etc., demonstrate a learned craft—speaks indeed for only the people shows. It is not necessary to expend another word on the subject, and that the people volunteer to come, is, of course, quite self evident."[150] Already aware that at least a few visitors to the shows found something in them which was basically demeaning, Leutemann justified the shows as, after all, just another form of entertainment in which someone puts him- or herself up on a stage to be looked at. The shows, he concludes, were in fact even more acceptable than the usual theatrical presentations because the people in the exhibits did not have to pretend that they were something that they were not, that is, they just "acted" naturally. Among other things, it should be clear at this point that although the shows were certainly a form of theatrical display, there is neither substantial reason to believe that the people in the shows "acted" naturally—or were perceived to have been what they really were—nor reason to accept Leutemann's assertion that the shows were no more degrading than any other form of theatrical exhibition. Again reflecting on the fates of the participants in his shows, Hagenbeck put the controversy down in a more matter-of-fact manner than Leutemann. He wondered, "Where are you all, you Africans, Indians, you red sons of the wilderness, you Eskimos and Laplanders who trusted yourselves to my leadership in the land of those remarkable Whites, who gazed at you in crowds as if you were fabulous animals."[151] "Fabulous animals," indeed. But this simile tells only part of the story. As much as the people in the shows were exhibited as animals—were put on display in the zoological gardens of Europe—it was Hagenbeck's experience with exhibiting people, his success in creating exhibits that made people believe that what they saw before them was nothing less than nature up close, which led this innovative thinker to change how we have come to imagine exotic animals. As we shall see, Hagenbeck's world of animals would soon come to resemble deeply the carefully constructed and reassuring exhibits that he had created for his indigenous peoples.

Coda

The last people show organized by the Hagenbecks arrived in Germany in the spring of 1931 from New Caledonia. Echoing Carl Hagenbeck's disastrous experience with a group of Indian servants brought from England in 1878 who were disturbingly more English than Indian, the house historian, Günter Niemeyer, reported that "even their arrival was disappointing. In their European clothes, they were hardly noticeable on the Hamburg streets, so that in all haste 'original costumes' had to be prepared for them based on examples in the

Ethnographic Museum." The problem seemed to be that in 1931, only two years before the Nazi seizure of power in Germany, it was very difficult to find "truly primitive" people who would fit into the established stereotypes; and Germany, it seems, had little interest in "civilized savages." Niemeyer complained that the boat that the "natives" were supposed to build with primitive tools capsized when it was launched and that, while "during the days they danced the obligatory Hula-Hula, in the evenings they jumped to the Foxtrott and Shimmy on the Reeperbahn."[152] The last people show was sent home early on June 21, 1931. But this is not really the end of the story of the people shows. The shows are still with us, although in many different forms. On the one hand they can be seen reasserting themselves in the passionate, memorable, and yet somehow disturbing photographs of the Nuba and the Kau made by Leni Riefenstahl—photographs that would serve to begin a craze of picture taking by tourists in Africa. On the other hand, echoes of the shows can be found in both late-twentieth-century documentaries of indigenous peoples and the modern "freak shows" airing as daytime talk shows on U.S. television. Sometimes, it seems, evidence of Hagenbeck's people shows can be found in what would seem the most unlikely of places. In a recent visit to the Holocaust Memorial Museum in Washington, D.C., for example, I found myself standing before a picture of Maggak, the woman exhibited with her husband and two children in the 1878 "Eskimo Show." In a display on Nazi racial ideology, designers had composed a wall of photographs of "racial types" which had presumably been taken from books on race written during the Third Reich. Maggak's picture, along with those of a number of other participants in the shows, is used in the exhibit. That exhibits of race would inevitably become exhibits of racism should not be too surprising at this point. In the racial texts, in Nazi ideology, in the museum exhibit, Maggak is a nameless "Eskimo." In Hagenbeck's exhibit, she was only a little more.

In variety theaters I have often watched, before my turn came on, a couple of acrobats performing on trapezes high in the roof. They swung themselves, they rocked to and fro, they sprang into the air, they floated into each other's arms, one hung by the hair from the teeth of the other. "And that too is human freedom," I thought, "self-controlled movement." What a mockery of holy Mother Nature!
—Red Peter, in Franz Kafka, "A Report to an Academy"

What made Hagenbeck's people shows such a success was their claim that they presented "exotic" people *as they really were*. In the construction of the often highly elaborate sets for these shows, for example, the company diligently sought to construct a convincingly realistic environment that would establish the authenticity of the exhibit. The goal was to re-create in Germany the native village of India, North Africa, Greenland, or the American plains, for example, and thereby seamlessly transport the exhibit's visitors to some faraway and yet highly imaginable locale. Looking back at some of the photographs taken of these exhibits, such as that of the "Bedouin Show" of 1912 (fig. 34), we can see that, especially within the frame of a photograph, the realism was often striking. Nevertheless, in the end these shows collapsed under the weight of their central conceit—that the image they presented was somehow *real*.

The eventual failure of the shows was due to a number of factors. First, the often hard-won optimism about the idea of "empire"—and this was especially the case in Germany, where the government and the people warmed very slowly to the prospect of German states in Africa, Asia, and the South Seas—became severely tarnished in the wake of native resistance and revolts in the 1890s and first decade of the twentieth century. Of course, whatever was left of the idea was all but demolished during and after World War I,

Fig. 34. "Bedouin Show" at Stellingen, 1912. From Alexander Sokolowsky, *Carl Hagenbeck und sein Werk* (Leipzig: E. Haberland, 1928).

an experience that taught the publics and governments of the European states a bleak lesson about the potential consequences of power politics. With the sheen of colonial conquest and expanded territories in places far from Europe almost irredeemably damaged, audiences to the people shows became increasingly critical about the worlds of order and happiness which the shows usually tried to portray. Second, with the increasing popularity of photographs and, especially, cinematic film as preferred methods to "study" foreigners—methods that allowed the viewer to examine the object relentlessly without ever being discomforted by looks coming back—the shows began to be displaced by anthropological films and popular geographical magazines. In confronting the older shows of people, these newer venues claimed an even higher level of authenticity—they claimed to show the genuine thing in a genuinely native environment. Finally, and perhaps most important, the Hagenbeck "illusion of the real" became impossible to maintain because the company simply could not find "genuine" natives; that is, the company could not find people willing to come to Europe who met the popular expectations for nativeness.

Propelled by the broad public's increasing skepticism about more traditional shows of people—shows that often had more in common with fairground freak shows than scientifically accepted ethnological displays—while building nevertheless on the public's fervent desire to see "authentic natives," Hagenbeck sought to create exhibits that would convince viewers that what they saw before them was not a bunch of performers in a fake village in the heart of a German city but rather a *real* village of *real* natives somewhere in a *real* exotic land. The "natives," however, simply didn't cooperate. They spoke up. They made it clear that they preferred to dress as Europeans while visiting Europe; they made it clear that they enjoyed alcohol, tobacco, and other amusements as much as their hosts did; they made it clear that they were able to learn European languages; and they made it clear that they were not shy about asking for tips and bartering with the Germans, French, and English *in* German, French, and English.

The tensions and contradictions inherent in the Hagenbeck people shows are also evident in Franz Kafka's narrative of the ape Red Peter, who is captured by Hagenbeck's agents. Recalling his first days on the steamer heading to Europe in the cage of three sides with the fourth made of the back of a locker, Red Peter reports, "All the time facing that locker—I should certainly have perished. Yet as far as Hagenbeck was concerned, the place for apes was in front of a locker—well then, I had to stop being an ape. A fine, clear train of thought, which I must have constructed somehow with my belly, since apes think with their bellies."[1] For Red Peter, becoming human proceeded in a series of clear steps. First he learned to spit, then to smoke a pipe. Though initially repulsed by the smell of schnapps, he finally overcame this obstacle as well. He explains:

> One evening before a large circle of spectators—perhaps there was a celebration of some kind, a gramophone was playing, an officer was circulating among the crew— when on this evening, just as no one was looking, I took hold of a schnapps bottle that had been carelessly left standing before my cage, uncorked it in the best style, while the company began to watch me with mounting attention, set it to my lips without hesitation, with no grimace, like a professional drinker, with rolling eyes and full throat, actually and truly drank it empty, then threw the bottle away, not this time in despair but as an artistic performer; forgot indeed, to rub my belly; but instead of that, because I could not help it, because my senses were reeling, called a brief and unmistakable "Hallo!" breaking into human speech, and with this outburst broke into the human community, and felt its echo: "Listen, he's talking!" like a caress over the whole of my sweat-drenched body.[2]

Turned over to his trainer in Hamburg, Red Peter advanced quickly on his path toward becoming a performer as a way of avoiding becoming once again a prisoner. He recounts: "I did not hesitate. I said to myself: do your utmost to get onto the variety stage; the Zoological Gardens means only a new cage; once there, you are done for."[3]

As I noted earlier, there is more to Red Peter's story than observations about human-animal interaction made by an ape who manages to become human. A central focus of the short story remains, for example, Red Peter's exploration of the nature of freedom. Certainly he ponders the freedom of an artist constrained by the expectations of a philistine audience, but also more generally he ponders the freedom imagined by anything feeling trapped. On one level, Red Peter suggests that freedom means escaping from a physical cage, and here Kafka is tapping into an already popular image of the zoological garden as little more than a prison for "innocent" animals. In an editorial in the *Daily News* (London) in 1869, for example, the commentator, speaking to an apparently sympathetic audience, wrote about a proposed lion house to be built at the London Zoo: "Lions at play, free as their own jungle home; tigers crouching, springing, gamboling, with as little restraint as the low plains of their native India—such is the dream of everyone interested in Zoology. We are all tired of the dismal menagerie cages. The cramped walk, the weary restless movement of the head . . . the bored look, the artificial habits. . . . Thousands upon thousands will be gratified to learn that a method of displaying lions and tigers, in what may be called by comparison a state of nature, is seriously contemplated at last."[4] In the end the New Lion House, completed in 1876 at the zoological gardens in Regent's Park, was far from the "state of nature" expressed in the editorial. It still consisted of heavily barred cages, although they were open to the sun and air in front and above. Nevertheless, more than anything else, it was those iron bars—marking so clearly the captivity of the animals—which repeatedly caught the attention of visitors to the zoos. The poet Rainer Maria Rilke, for example, stood before the panther's cage at the Parisian zoo, the Jardin des Plantes, in 1907 and concluded that the bars must constitute the animal's entire reality: "The bars which pass and strike across his gaze / have stunned his sight: the eyes have lost their hold. / To him it seems there are a thousand bars, / a thousand bars and nothing else. No world."[5] Indeed, for many observers at the end of the nineteenth century, the zoo was, before all else, a place of captivity, a place where animals were locked up.

On another level, however, although Red Peter was desperate to find a "way out" of the iron cages, he distinguishes that desire from a wish for "freedom." He explains: "I fear that perhaps you do not quite understand what I mean by 'way out.' I use the expression in its fullest and most popular sense. I deliberately do not use the word 'freedom.' I do not mean the spacious feeling of freedom on all sides. As an ape, perhaps I knew that, and I have met men who yearn for it. But for my part I desired such freedom neither then nor now. In passing: may I say that all too often men are betrayed by the word freedom. And as freedom is counted among the most sublime feelings, so the corresponding disillusionment can be also sublime."[6] Red Peter's point is that what humans call "freedom"—the movement, according to his example, of trapeze artists floating through the air—is not the freedom he knew as an ape. That is only a highly constrained illusion of freedom. In human society, he argues, there can be no true freedom, no true "spacious feeling," but only brief, cramped imitations or performances of it. These were the performances, of course, which were given at the people shows, and these were the performances Carl Hagenbeck's nonhuman animals were soon to give.

First in his experimentation with new techniques for training animals for performance and then in the creation of his "zoological paradise" in Stellingen, Carl Hagenbeck attempted to resolve the tensions behind the first level of the critique voiced by Red Peter—the imprisonment of innocents in iron cages. In the end, Hagenbeck's remarkably effective response to that critique fundamentally changed the basis for the exhibition of wild animals in domestic spaces ever since. Creating what became known as barless enclosures—a way of isolating animals from one another and the public through the use of landscape elements such as moats and rock outcroppings—Hagenbeck set a standard for exhibiting animals which can be seen today in the most modern exhibits in contemporary zoological gardens and aquariums. Hagenbeck's response to the issue of captivity, however, was—to bring back Red Peter's second level of critique—necessarily circumscribed by the illusory quality of the "freedom" that the animals in Hagenbeck's paradise enjoyed. Indeed, the deep connection between Hagenbeck's exhibits of people and his later exhibits of animals will lead us at the close of this study to a consideration of the fundamentally parallel underlying dilemmas flowing through both kinds of exhibit. This examination will show, in the end, the severe limitations of what we might expect from zoological gardens in the twenty-first century.

The Circus: "A Humanely Trained Lion Act"

Carl Hagenbeck's first step in rethinking how animals were exhibited began with his experiments in putting together circus acts. Importantly, these experiments grew directly out of the shows of people. According to Hagenbeck, a perplexing legacy of one of his"Ceylon" shows was a large herd of elephants with no immediate buyer. Stuck with the elephants, the equipment from the show, and a large assortment of artifacts, and having already visited the large exhibition halls of Europe, Hagenbeck hit upon the idea of organizing a traveling big-top circus after the American fashion. The large tent, he thought, would give the show a new flexibility and allow the old exhibit to travel to cities for which such a show would be quite a new event. Thus, with a group of twenty artistes from Sri Lanka, a range of other performers, groups of trained animals, and a herd of working elephants, "Carl Hagenbeck's International Circus & Singhalese-Caravan" opened on the large area known as the Heiligengeistfeld in Hamburg in 1887. As in the case of the people shows, Hagenbeck noted that the idea to form a traveling circus had more to do with an immediate financial consideration than anything else: he argued, "I plunged right into the new undertaking, but I was responding more to necessity than a personal desire, because, I must admit, I did not have a deep interest in this business."[7] Unlike the people shows, however, although the Hagenbeck circus did bridge a temporary financial problem and the company continued almost uninterruptedly to train animals for performance through to the beginning of World War I, Hagenbeck, we are told, never acquired a passion for the circus life. His son Lorenz, for example, recalled that his father repeatedly pointed out to the young circus enthusiast, "No, my child, you will not become a gypsy!"[8]

In the end, Hagenbeck's first venture with the world of the circus was, it seems, simply born under an unlucky star. On the opening day in Hamburg, a storm blasted the tent to pieces, and Hagenbeck himself was nearly crushed by a falling tent support. A new tent was ordered, however, and the show completed two tours of Europe. Nevertheless, Hagenbeck was continually frustrated by what he called the "rabble" associated with the circus world and its performers. Writing in his memoirs in 1908—shortly after running into legal difficulties with his American partners in another circus enterprise that eventually led to the establishment of the American Hagenbeck-Wallace Circus[9]—Hagenbeck simply concluded, "With these people it is difficult to come out even; fidelity and trust do not live in their hearts."[10] After repeated disasters, including an episode in Munich in 1888 in which eight of his elephants panicked during a parade and stormed through the city, injuring hundreds of people and killing several, Hagenbeck sold the properties of his first circus in 1889.

The story of Carl Hagenbeck's involvement with circuses does not end with the closure of "Carl Hagenbeck's International Circus & Singhalese-Caravan," however. Indeed, in the same year Hagenbeck, with the help of his brother Wihelm, who had his own animal training school, had debuted his newest attraction—a group of four lions trained, he claimed, in an absolutely new way that put aside, in his words, the "old, cruel methods of training animals."[11] Over the years, he argued, he had arrived at the conviction that "animals are creatures like us and their intelligence is different from ours only in degree and strength, but not in type. They react to meanness with meanness, and to friendship with friendship." Hagenbeck concluded that "through love, kindness, and perseverance, paired with discipline, one can get more out of an animal than through raw force."[12] Hagenbeck wanted to do away with the whip, club, red-hot irons, pistols, and metal bars by carefully selecting animals that seemed to have an aptitude for learning tricks, training them through praise and bits of meat and only occasionally reprimanding them with a whip when the animals became sloppy or careless.

According to Hagenbeck, between 1887 and 1889 he worked with more than twenty lions before settling on four for intensive training. These lions were trained to perform a series of remarkable acts, at the close of which, Hagenbeck noted, "the tamer drove a two-wheeled cart—resembling the form of an ancient Roman chariot—which was pulled by three lions four times in full gallop around the forty-foot across circular cage—a sensational number."[13] From 1889 to 1892 the lions toured Europe with great success while Hagenbeck and his assistants continued training other groups. More than the traveling circus, it seems, the lions were a highly profitable undertaking, and, Hagenbeck argued, the act became one of the most lucrative activities of his whole business.[14]

The success of the lions encouraged Hagenbeck to train more groups of large animals and to begin to prepare a large show for the World's Columbian Exposition in Chicago in 1893. What began as an exciting idea, however, ran into repeated difficulties. First, Hagenbeck's premier group of performing animals—including twelve lions, two tigers, a cheetah, two South American sun bears, and a polar bear—all died of poisoning from spoiled meat they had eaten during their stay at the Crystal Palace in 1891. Then, when Hagenbeck tried desperately to replace the animals with new shipments of tigers and lions, every one of the young animals arriving in the spring and summer of 1892—at least seven tigers and four lion cubs—died of what Hagenbeck believed was cholera, which began to devastate the human population of Hamburg later that summer. According to Hagenbeck, within a few months he had lost some 70,000 marks' worth of animals. Still, with a reserve trained group of lions, tigers, and bears, as well as some 150 primates and a collection of 80 different species of parrots, Hagenbeck arrived in Chicago in the spring of 1893. Unfortunately, the huge building that he and his American partners were constructing on the midway at a cost of more than $100,000, which was modeled, it was said, on the Coliseum in Rome and designed to accommodate 40,000 people and seat 6,000 for the performances, was not yet complete, and before the exhibition opened, Hagenbeck lost an additional $2,000 worth of animals to inclement weather and poor conditions.

Despite the difficulties, accounts of the exposition relate that the Hagenbeck Pavilion provided stunning entertainment. According to the souvenir album, *The World's Fair, being a Pictorial History of the Columbian Exposition,* for example: "Among other features of the performance, there is a revival of the grand gladiatorial contests of the Empire, under supervision of the Director of the Roman National Theatre. All of the most interesting features from Rome's most glorious epoch are illustrated in actual life. There are no flimsy theatrical effects. Large numbers of well-built and powerful men, as well as a corresponding number of the handsomest Italian women, all belonging to the peasantry, participate in these performances. There are combats between men and wild beasts, chariot races and other contests." The show—much in the spirit, it seems, of one of Hagenbeck's people shows but here showing a historical spectacle— continued with a triumphal return of an army, complete with male and female prisoners bound for slavery. "Bacchanalian dances" and a scene of the persecution of Christians were evidently presented, "with an impartiality that [would] satisfy the most critical spectators."[15] In all, the exhibit, with its appearances by emperors and senators as well as common street people, was intended to give a sense of life in ancient Rome. The guide book, however, then turned to the animal exhibition: "But the most fascinating part of the exhibition is taken by the animals, some of whose performances are simply marvellous. Bears walking the tight rope and doing the William Tell act, as well as ermine-mantled and crowned lions driving triumphal chariots around the arena, drawn by royal tigers, may be seen. Camels hump themselves on roller skates. The hippopotamus essays an act on the trapeze, while the untamable rhinoceros appears as a tractable actor." The animal portion of the program continued with "hypnotized monkeys ready to converse with any intelligent visitors in their own language," parrots playing "progressive euchre" which "'differ' about the prizes in sixty-five different languages," and a dwarf elephant named Lily, which, although some ten years old, was only about three feet high and three-and-a-half feet long.[16] Hagenbeck's arena was clearly a place that promised entertainment (fig. 35).

Beyond this general description, however, it is possible to be a good deal more specific about the precise nature of the animal exhibit at Hagenbeck's

Fig. 35. The Hagenbeck Pavilion, World's Columbian Exposition, 1893. Courtesy, Fred Pfening Collection.

Fig. 36. World's Columbian Exposition: Cover of official program, *Hagenbeck's Arena and World's Museum*, 1893. Chicago Historical Society.

arena. A surviving official souvenir program for the show, *Hagenbeck's Arena and World's Museum*, picturing on the cover Carl Hagenbeck and his brother-in-law and animal trainer Heinrich Mehrmann (fig. 36), makes clear that the show was broken into two parts taking place in an innovative circular cage. The first consisted of a series of diverse animal acts: Marcella Berg presented an assorted group consisting of an elephant, two Shetland ponies, and two boar-hounds. Then a Mr. M. Beketow, a Russian clown and trainer, performed with his "Trained Pigs and Wild Boar." After Beketow, the "dwarf elephant Lily"—"The Smallest Elephant in the World: The most remarkable Zoological Curiosity of the nineteenth Century"—was introduced. Then came one of Wilhelm and Carl Hagenbeck's signature acts, "The Equestrian Lion," in which one of three lions rode a horse around the arena under the guidance of either Ella Johnston, William Philadelphia, or John Penje.

The second part of the performance featured the trainer Heinrich Mehrmann. His introduction in the program captures a sense of the stupendous, which was, of course, very much a part of these events. The text reads: "The King of Animal Trainers presenting a constellation of Trained Animals, consisting of Lions, Tigers, Panthers, Leopards, Boarhounds, Polar, Sloth and Thibet Bears, all in the same arena and going through a performance the possibility of which can only be realized when witnessed. The degree of perfection to which these Wild Animals are trained is almost past belief. This performance has been patronized by all the crowned heads of Europe." Mehrmann's performance consisted of twelve set pieces, which, building one upon the other to the triumphal finish, are worth a complete listing: "1. The Bear on the Barrel; 2. Tigers on the Tricycle; 3. Lions on the Tricycle; 4. The Bear on the Tight Rope; 5. The Hurdle Race—Boarhounds leaping over Hurdles which are supported by Lions and Tigers; 6. The See Saw—Performed by Lions, Tigers, Panthers and Bears; 7. The Zoological Staircase—comprising Lions, Tigers, and Boarhounds; 8. The Bengal Tiger on the Rolling Globe; 9. Bear on the Rolling Globe—Walking and rolling same standing on his hind legs; 10. Steeple Chase of the Boarhounds—Leaping over living hurdles of Lions and Tigers; 11. Triumphant Drive of the Lion Prince—Dressed in Royal Robes and Crowned, drawn in his chariot by two Bengal Tigers and having two Boarhounds as footmen; 12. The Great Zoological Pyramid—Consisting of Lions, Tigers, Panthers, Leopards, Bears, and Dogs."[17] What seems to have repeatedly struck visitors to the Hagenbeck arena, especially in such acts as the evidently startling "Drive of the Lion Prince" (fig. 37), were the performances in which diverse animals, though popularly conceived to be mortal enemies, performed together with apparently no enmity between them. As one extravagantly produced souvenir album of the exposition explained under a photograph of a lion beside a dog titled "The Happy Family":

The lessons taught at Hagenback's [sic] menagerie were far outside the ordinary course of study. They embraced illustrations of the power of human kindness and the possibilities of the redemption of animals from their state of ferocity; they proved how the dog (if not the lamb) might lie down with the lion and neither be the worse for the experiment. The illustration exhibits the mutual confidence, the banker-depositor faith which existed between the "King of the forest"—and there were several of them—and the jolly canine of the arena, the cheerful dog who thought nothing of nipping a lion by the leg or whispering in his ear with his teeth. As for the mutual trust in each other, that is shown by the droll, I'm-up-to-snuff look of the dog, and the apathetic waiting-to-hear-what-you-say demeanor of his companions and associates. These little assemblages *en banc,* as it is said of the conventions of judges, were the most delightful exhibits the children could obtain on the Midway, and no matter how old those boy and girl auditors may become, they never can or will forget the sensationalism of these continually renewed treaties of peace between man and brute, and brute and man, as given in Hagenback's Arena.[18]

Indeed, the dominant motifs of this passage—including the lions lying down with other animals and the "illustrations of the power of human kindness"— recur throughout reports of Hagenbeck's exhibit in Chicago and clearly represent the focus of the broad public appeal that the show garnered. As a German picture album explained, "Hagenbeck's menagerie proves the fulfillment of the prophecy in that inside the building the lion lies down beside the lamb and the most defenseless animals, while they couldn't possibly protect themselves, play with all surety with the most vicious beasts. The presentations of the well-trained animals were watched with wonder by thousands of visitors every day.

Fig. 37. "Drive of the Lion Prince," World's Columbian Exposition, 1893. From *Midway Types* (n. pub., n.d.). Chicago Historical Society.

Fig. 38. Poster for Hagenbeck's Trained Animals, World's Columbian Exposition, 1893. Courtesy, Ken Harck Collection.

Herr Hagenbeck is known around the world as the most successful teacher and tamer of wild animals. He illustrates—as no other living person can—the power of the human spirit over the animal's mind and brute strength."[19] Looking back at such descriptions—and looking back as well at one of the stunning posters of Hagenbeck's exhibit in Chicago (fig. 38)—it is clear that the Hagenbeck exhibit represents a significant milestone in the exhibition of performing animals. While it is true that Hagenbeck had been showing the results of his experiments for several years, it was the performances of his animals in Chicago in front of the at least one million visitors to the Hagenbeck Pavilion—and over the next year in New York at Madison Square Garden, Manhattan Beach, and then back at the Garden—which established his name as one of the great figures in the history of animal training.[20]

For his part, Hagenbeck claimed that his most important innovation lay in his new method for training animals according to their aptitude and through kindness, and it has become something of a habit in the Hagenbeck histories, as well as in histories of circuses, to credit Hagenbeck with the introduction of humane methods to the training of performance animals, especially the big cats. Although this assertion is partly true, the statement demands qualification. First of all, it is also true that not all animal tamers—despite their popular depictions in late-nineteenth- and early-twentieth-century dime novels—used brute force and intimidation to train their charges. That a training manual from 1869, for example, could point out that much more can be achieved through training animals through kindness than through violence makes clear that Hagenbeck's ideas were not completely revolutionary. According to the manual, *Haney's Art of Training Animals,*

> when [an animal] knows what you want him to do he will in almost all cases comply with your wishes promptly and cheerfully. For this reason punishments seldom do any good, unless the animal is willful, which is rare. On the contrary they, as a general rule, interfere with the success of the lessons. If the pupil is in constant fear of blows his attention will be diverted from the lesson, he will dread making any attempt to obey for fear of failure, and he will have a sneaking look which will detract materially from the appearance of his performance. This is the case with the animals instructed by a trainer of this city who "trains his horses with a club," the animals never appearing as well as those taught by more gentle means. But for a rare natural talent this man's success would have been utterly defeated by his brutality. He is the only one we know of in the profession who does not base his tuition on kindness to the pupil. A sharp word or a slight tap with a small switch will as effectually show your displeasure as the most severe blows. It is both cruel and unwise to inflict needless pain.[21]

In addition to a broader consensus about training through kindness than Hagenbeck might have us believe, by 1890 it was also apparent to most figures within the circus world that the public was becoming increasingly uncomfortable with more violent methods of training and presenting the larger, more dangerous animals. Concurrently with Hagenbeck, for example, Frank Bostock and the brothers Francis and Joseph Ferari, all from England, were also employing more humane methods in training their animals and to equal effect. Indeed, that the circus world was becoming very sensitive to criticism of "cruel" training methods is clear in the 1903 official brochure of the Barnum and Bailey Circus: "One thing more concerning these exhibits of trained animals of all kinds in this show may be fairly and truthfully said. They are trained by kindness. Mr. J. A. Bailey, the controller and operator of the Greatest Show on Earth, is one of the foremost members and supporters of the Society for the Prevention of

Cruelty to Animals. He most rigorously prohibits any exhibit of temper on the part of the trainer either in public or private."[22]

Second, just as it is not entirely accurate to claim that Hagenbeck came up with his ideas of training animals out of some kind of vacuum, it is also not completely true to say that his trainers did not find themselves in a profession that often required a good deal of what might well have been called brutal training methods. In an 1883 *Forest and Stream* article, for example, a J.S.W., who claimed to have studied with Hagenbeck in Hamburg, argued that intimidation was the stock and trade of the animal trainer and that "the whip [was] the trainer's chief reliance; and it [could not] be used too freely."[23] Similarly, William Philadelphia, who learned Carl and Wilhelm Hagenbeck's techniques in Hamburg and worked in the ring for Carl Hagenbeck and Heinrich Mehrmann in Chicago and New York in 1893 and 1894, recalled his training of the famous lion "Black Prince" in ways that do not really stand up to Hagenbeck's assertions about using only those animals that show an aptitude for learning and never using force to train them. Philadelphia, who had been charged with the task of raising and taming the cub, which had recently arrived from Africa, described this encounter after days of patient talking to the animal while standing outside its cage: "I went into the cage, carrying a wooden club in each hand. The first time I entered he sprang at my throat, as his fierce instinct taught him to do. I gave him two or three good raps over the head and flanks, and he went back, not roaring, but making the queer-sounding purr peculiar to young lions. Then he came at me again and again. I used my clubs, but not too hard, avoiding hurting him badly, and being careful not to strike him on the back, for a young lion's back is easily broken. After feeling the club several times he kept away from me, and went into a corner of the cage sulking."[24] After weeks of such training—the retelling of which must have brought back to many *McClure's Magazine* readers the similar beatings suffered by the dog "Buck" in Jack London's *Call of the Wild*—Philadelphia noted, "The wild beast is afraid, not so much of any pain which may be inflicted upon him, but of some vague, unknown power too great for him to understand or cope with. This is what gives the tamer his control of lions and tigers. It is not any personal magnetism or any inherent virtue not possessed by other men. It is no charm of the eye. That idea is clear nonsense. It is merely that one particular man, by untiring patience, has succeeded in making himself appear in the lion's eyes as the one great and boundless force of the universe before which he must bow."[25] To a certain degree, of course, Philadelphia's account should be understood as part of his effort to make a powerful impression of himself to the readers of *McClure's*—the idea of the brutal animal trainer still had a certain romantic appeal. At the same time, however, it seems fairly clear from the ease of his descriptions that the beating of lions, tigers, and bears was still a widely accepted part of training regimens.

While Hagenbeck—and his chroniclers and historians—have repeatedly exaggerated the revolutionary nature of his "new methods" in training animals, however, he—and they—have almost completely ignored the monumental change that his performing animal troupes in fact wrought in the history of animal exhibition. To understand that change, we need to look beyond issues of training and focus instead on the nature of the performances of Hagenbeck's groups. Just how different they were from virtually all those that preceded them is clear, for example, when we briefly compare them with the spectacles created by Isaac A. Van Amburgh, arguably the most popular and influential lion tamer of the sixty years between 1833, when he first stepped into a lion's cage, and 1893, when Hagenbeck's show appeared in Chicago.

Van Amburgh's act appears fairly simple, and its central qualities can be easily recognized in the portrait by the artist Edwin Landseer (fig. 39). Dressed as a Roman gladiator, the brave performer would enter a small rectangular cage of lions, tigers, and leopards. Through demonstrations of his physical and what he called "moral powers," he would then enter into combat with the ferocious beasts, which were soon overwhelmed by "the natural beauty of the human form" and by the tamer's trademark crowbar, which was wielded with impunity.[26] The violent ambience of this exhibit is suggested well in an account of Queen Victoria's visit to Van Amburgh's performance at the Theatre Royal, Drury Lane, on January 24, 1839. The account was repeatedly produced in publicity for Van Amburgh throughout his career.

> On this latter evening . . . the Queen condescended to cross the stage of the theatre for the purpose of seeing the animals in their more excited and savage state, during the operation of feeding them. This gratifying scene took place after the departure of the audience, and every precaution [was] taken for the comparative comfort and seclusion of the Royal visitor, which the resources of the theatre permitted. The animals had been kept purposely without food for six-and-thirty hours, strong symptoms of which had became [*sic*] manifest during Mr. Van Amburgh's performance, by the lion and the panther having simultaneously attacked the lamb on its

being placed in their den; and they would evidently [have] made but a mouthful apiece of it, had not their almost super-human master literally lashed them into the most abject and crouching submission. The first portion of food thrown amongst them, seized by the lion as a matter of priority, was enough to convince any skeptic of the fearful savageness of their nature, when out of control of the one hand whose authority they acknowledge. The rolling of the tiger's eye, while he was devouring the massive lump of meat and bone, clutched between his fore-paws, seemed to possess the brilliancy as well as the rapidity of lightning; and was only diverted by a tremendous and sudden spring of the lion, who, having demolished his own portion, seized upon what was left of his ferocious neighbor's fare. The dash against the sides of the den sounded like the felling of huge trees, and was enough by its force and fury to shake the strongest nerves; but it was a positive fact, that while the boldest of the hearts in the royal suite speedily retreated at this unexpected plunge of the forest monarch, the youthful queen never moved either face or foot, but with look undiverted, and still more deeply riveted, continued to gaze on the novel and moving spectacle.[27]

Fig. 39. Edwin Landseer, *Portrait of Mr. Van Amburgh, as He Appeared with His Animals at the London Theatres*, 1847. Yale Center for British Art, Paul Mellon Collection.

Fig. 40. "A Dangerous Friend"—Tiger on the Tricycle, World's Columbian Exposition, 1893. From *Midway Types* (n. pub., n.d.). Chicago Historical Society.

Following in the footsteps of Van Amburgh's performances—in which indeed the lion did lay down with the lamb, but only after being beaten into submission[28]—came an entire series of lion tamers whose goal was to create exhibits that would yield the impression that the tamer's life was in jeopardy at every moment.[29] That a good number of tamers in fact and in fiction died in these sorts of spectacles only seemed to increase their popularity. Even the performances of John "Grizzly" Adams, called in early articles the "Californian Van Amburgh," which featured the abiding friendship of Adams and his bear, "Benjamin Franklin," eventually came to focus on the staging of deadly fights between Adams and assorted bears with "Ben" fighting at his side.[30]

To be sure, Hagenbeck's trainers were also frequently characterized as pitting their lives against potentially violent beasts. Nevertheless, when we consider the Hagenbeck performances, with such acts as those shown in the poster of the "Tiger on the Tricycle" (fig. 40), "The Great Zoological Pyramid" (fig. 38), and "The Drive of the Lion Prince" (fig. 37), it is clear that the ideas advanced by Hagenbeck's performances represent a very new way of imagining animal-animal and animal-human interaction. Instead of a gladiator beating the beasts into submission, here we see a former Hamburg businessman, Heinrich Mehrmann, as he stands respectfully in evening wear beside his calm, far-from-roaring charges. As a reporter writing for *McClure's Magazine* noted, "One of the most remarkable things that I noticed in Karl Hagenbeck's menagerie is the marvellous unity and loving-kindness which is brought to pass amongst his animals. They are fondling and playing with each other the whole day long."[31] Reviewing the show in Chicago, another writer could only wonder optimistically, "Perhaps history is at fault, and there are grievous errors about human sacrifices to the lions: or perhaps the lions or the kind of men are not the same as two thousand years ago; or, best perhaps of all, the Exposition may have established bonds of peace and unity between man and beast."[32] What was truly remarkable about Hagenbeck's performing animal exhibits was that they tied deeply into a desire in the audience for a highly sentimentalized view of animals both in nature and in captivity. The time-honed contention that he trained his animals through kindness is really beside the point. That argument itself is only one more manifestation—and is perhaps even a necessary corollary—of an overall image Hagenbeck attempted to convey in his exhibits of performing animals. That image, which fully exploited ideas of the humane trainer walking with confidence among his pupil animals—animals that respect, admire, but more than anything love their keeper—came to its eventual apotheosis in Hagenbeck's groundbreaking Animal Park.

The Animal Park: "A Paradise for the Animals"

In the last decade of the nineteenth century, Hagenbeck found himself needing more and more space to house burgeoning stocks of animals. On the one hand, the rapidly growing company was receiving orders from new zoos all over the world including those in Africa, Japan, China, and South America. Meanwhile, the market for exotic animals among private individuals—including such figures as the Russian Frederic von Falz-Fein, who became an important figure in the preservation of the Przewalski's horse; the duke of Bedford; and Baron Walter von Rothschild—was also expanding. Some of these private collections, moreover, rivaled any public zoological garden. In addition, hunting parks were increasingly being stocked with Hagenbeck's animals, and Hagenbeck himself began to suggest to prospective investors—correctly foreseeing what would eventually come to exist in places around the world almost a century later—that it might be possible to build large private game parks in Europe and America where one could hunt bears, moose, elk, and deer, as well as such animals as lions, tigers, and elephants.[33] Whereas in the 1880s, Hagenbeck sold perhaps twenty game animals in a year, by the 1890s that figure had jumped into the hundreds. In any case, with the increasing volume of animals passing through Hagenbeck's property on the Neuenpferdmarkt—at one point there were twenty elephants there—it became clear that the company would have to move to larger quarters.

Through an acquaintance in the Prussian village of Stellingen, today a suburb of Hamburg, Hagenbeck heard that property in the area was for sale. It quickly became apparent that, indeed, the fields surrounding the village were ample to meet Hagenbeck's unusual needs, and, more important, large areas were available for purchase. Within a couple of days, Hagenbeck had acquired two hundred thousand square feet for thirty-five thousand marks. Within a few more days, two adjoining parcels of land were obtained, and within five months, twelve large enclosures had been arranged on the property. Finally, with the financial backing of a small consortium, Hagenbeck managed to procure all the property between Stellingen and the border of Hamburg at Elmsbüttel, allowing him eventually to build a railway spur giving direct access to Hamburg. In October 1902, when all the purchases had been made and Hagenbeck controlled some fourteen hectares of land for his park, the building of his new Animal Park began.[34] After constructing artificial mountains and ponds and moving some 40,000 cubic meters (more than a million cubic feet) of soil, Hagenbeck, his architect, the Swiss Urs Eggenschwyler, and the landscape architect H. Hinsch converted a field with six trees into what Hagenbeck described as a scene from the Arabian Nights.[35]

Hagenbeck argued that his objectives with the new park were fairly simple: "The leading thought was to present the animals in the most freedom and thereby demonstrate at the same time what acclimatization was able to accomplish. I wanted to show animal lovers with a large, practical and lasting example, that it is absolutely unnecessary to construct luxurious and costly buildings with large heating systems in order to keep animals alive and healthy. On the contrary, having the animals reside in the outside air and become used to the climate presents a far better method for protecting their lives."[36] But on May 1, 1907, when Hagenbeck's new Animal Park in Stellingen opened to the public, it became immediately clear that while Hagenbeck's simple goal was to prove that animals of the tropics did not need to be kept in expensive, hot, and humid buildings resembling elaborate greenhouses, the entrepreneur's new facility represented an utterly new vision of how exotic animals could be kept and, more important, how they might be displayed. Hagenbeck called his creation an "animal paradise," relating in his mind—and for the public—his park and the biblical Paradise, a place where "animals would live beside each other in harmony and where the fight for survival would be eliminated."[37]

Using data that he had begun to collect in the 1870s when he first began to experiment with training animals, Hagenbeck had established the vertical and horizontal jumping abilities of a large number of animals. As usual, his tests were simple and practical: in one trial, for example, he tied a stuffed pigeon to the top of a pole and measured the leaps of the great cats in their attempts to procure the bird. From this statistical material, Hagenbeck designed enclosures for his animals without the bars and fences that were presumed to be the "necessary" tools of the zoological gardens of the period. In their stead, as we can see from a photograph of the "Enclosure for Grazing Animals" (fig. 41), Hagenbeck developed systems of dry and water-filled moats that were either concealed with vegetation or employed as "natural" ponds and used them to separate the animals from one another and the public. But even more innovatively, Hagenbeck arranged his gardens so that numerous discrete enclosures could be observed at once. From a single viewing point, one looked out across a small body of water to an apparent riverbank enjoyed by ducks, flamingos, and small deer. Beyond them, larger antelopes, ostriches, and zebras milled about, seemingly watched from still farther back by lions resting beside water holes or in the shade of "rock" grottos. Dominating the entire panorama, a rugged "cliff" provided habitat for wild goats and vultures. "In the Stellingen animal park," wrote Hagenbeck, "one could observe for the first time never-before-seen tableaus, like African ostriches during the winter taking snow-baths in the open air or romping about in pleasure in even colder weather." With his designs, "ibexes, chamois, and antelopes need not trust their lives in captivity to low cages, but rather could strive for the heights on a cliff-like ridge. The king of the animals moved about in freedom, in proud majesty in his wide grotto."[38]

Indeed, Hagenbeck's park was not a zoological garden in the sense in which that term was generally understood in the nineteenth and early twentieth cen-

Fig. 41. Hagenbeck's Animal Park, enclosure for grazing animals, mid-1920s. From Alexander Sokolowsky, *Carl Hagenbeck und sein Werk* (Leipzig: E. Haberland, 1928). Courtesy of Hagenbeck's Tierpark.

turies. Most nineteenth-century zoological gardens were founded through associations of public and private interests which sought to present the world's wildlife in an appealing setting, with the objective of furthering both scientific and popular knowledge of animals. In contrast, Hagenbeck's Animal Park in Stellingen was an integral part of Hagenbeck's more general animal business. As he noted, "My animal business had taken on ever greater dimensions. . . . It was like a great tree with many branches and twigs. From the original trunk grew the people shows, the animal training, the attempts at domestication and breeding, and many more developing projects continued to struggle to reach their form and flower."[39] His 1907 park in Stellingen was, Hagenbeck argued, the extension or perhaps the conclusion of almost sixty years of the company's growth. As much as the Animal Park might seem to be the simple and logical extension of a company needing room to expand its animal trade and present its people shows and groups of performing animals, however, a closer look at the roots and legacy of the park makes clear that its creation represents a pivotal moment in a modern reorientation in thinking about the lives of animals in captivity.

Like a great many of Carl Hagenbeck's innovations, the roots of the Animal Park should not be traced as much to Hagenbeck's wanting to change the way people thought as to his wanting to make some change for company. More than a decade before the park opened, Hagenbeck had debuted his new kind of animal exhibit to enthusiastic reviews—his exhibits made money as well as influenced people. An 1898 poster advertising one of these new attractions, "Carl Hagenbeck's Zoological Paradise—The Zoological Garden of the Future," made clear just how novel these exhibits were. The poster quoted the instructions of the directors of the Berlin Zoological Gardens to those visiting his attraction in their city:

> One should imagine that one finds oneself transported by means of a magic wand to the Paradise of the Bible when entering into Hagenbeck's Zoological Panorama installed in our Zoological Garden near the entrance to the metropolitan railway. All that is missing to complete the illusion is the presence of Adam and Eve! In a picturesque landscape strewn with rocks, all sorts of animals are seen walking around, not behind wire fencing, as one would be tempted to believe, but in complete freedom! We see llamas, alpacas and guanacos, bulls and cows, goats and ewes, deer, antelope, wild donkeys and zebras, camels, elephants and kangaroos, cranes and storks, peacocks, swans, geese and ducks, gulls, etc., running and gamboling. But it is not just a matter of these not very dangerous animals; we see there, equally at liberty, all kinds of savage animals: lions and tigers, leopards and pumas, and all manner of bears and hyenas, at the sight of which many spectators will have felt at first a desire to retrace their steps rather than advance.[40]

Despite the apparent danger, however, the advertisement counsels the public not to fear: "This paradisal peace . . . has been forcibly achieved by very modern and practical means, means absolutely secure which have been invented by Mr. Carl Hagenbeck himself, and which have even been patented in Germany."[41] Through Hagenbeck's plans, the advertisement continues, "the savage animals find themselves 'by chance' always in the background of the panorama, and don't even think, it appears, to give themselves the trouble to change their ordinary menu through a succulent roast, of which they seem only to have an embarrassment of choice amid all that living flesh which abounds a few steps from them."[42]

Drawn for the Hagenbeck exhibit at the Berlin Zoological Gardens, the illustration of Hagenbeck's zoological garden of the future (fig. 42) presents what must have seemed to contemporary viewers an impossible scene.[43] At the bot-

Fig. 42. *Carl Hagenbeck's Zoological Paradise, The Zoological Garden of the Future*, 1898.
Courtesy of Hagenbeck's Tierpark.

tom of the work, a family looks out over a railing at a landscape filled with animals. In the foreground, pelicans, flamingos, marabou storks, swans, cranes, geese, an ibis, and a white stork surround and swim in a small pool. Behind them, various kinds of antelopes, a dromedary and a Bactrian camel, several kangaroos, a llama and an alpaca, ostriches, assorted deer, a tapir, a water buffalo, a rather stuffed-looking elephant, and a scrawny zebra stand around a crowded open area bordered on the sides by large trees. Beyond them, in somewhat more rugged territory, all manner of carnivores lie and walk about—most of them attentively observing the viewers. Among the rocks one can make out four or five tigers, five lions, a hyena, three small performing bears begging from a caretaker, and three polar bears playing on a log. In the distance beyond these carnivores, mountains rise up, inhabited by a pair of ibex and a creature that appears to be a yak.

Hagenbeck's traveling "Zoological Paradise," as it was installed in the Berlin Zoological Gardens, presents what proved to be the predominant theme behind his later, more permanent constructions. With minor changes, in fact, the illustration for the traveling exhibit could just as well have served to illustrate an advertisement for the main panorama of the later Animal Park (fig. 43). The excitement over Hagenbeck's panoramas, however, traces back to two years before the "Zoological Paradise" to his first animal panorama, which appeared in the form of an Arctic landscape in 1896. This panorama, shown in Hamburg on the Heiligengeistfeld, then in Berlin at the Commercial Exposition of that year, and later in Dresden and Paris, had taken advantage of a particular historical moment with almost unimaginable success. In a year in which Europe was riveted by the Norwegian Fridtjof Nansen's exploration of the Arctic in his ship, the *Fram*, Hagenbeck designed a traveling animal exhibit that presented an Arctic world filled with seals, sea lions, polar bears, and even Antarctic animals such as penguins. The animals were seen swimming in pools and walking about a landscape of artistically rendered ice and snow. At the right rear of the scene, however, Hagenbeck presented a replica of the surging bow of the *Fram* frozen in ice.[44]

Of course, even the *Fram* exhibit had its precursors. It bore, for example, a relation both to a longer tradition of entertainment within the firm and to a contemporary popular enthusiasm for historical panoramas. In order to appreciate just how the animal panoramas represented something quite new, however, we must briefly examine these other influences. With respect to the longer his-

Fig. 43. Hagenbeck's Animal Park, main panorama, mid-1920s. From Alexander Sokolowsky, *Carl Hagenbeck und Sein Werk* (Leipzig: E. Haberland, 1928). Courtesy of
Hagenbeck's Tierpark.

tory of the firm, it is helpful to reconsider perhaps the most repeated story of
the Hagenbeck firm, that of the exhibition of the six seals in 1848.[45] Carl Hagenbeck's brief account of the beginning of the family's exotic animal business
can be read over and over again with little variation in the company's official
histories. John Hagenbeck's biographer, Wilhelm Munnecke, however, described the exhibit in Berlin in somewhat fuller detail. In an account that contrasts sharply with the story told by Carl Hagenbeck, Munnecke writes: "With
these six seals, [John Hagenbeck's] father laid the cornerstone to his wealth.
He knew, or at least felt, that a handful of money was to be made with the
curiosity of people. On a site in Berlin he had put up a booth over the entrance
of which was a sign with the inscription: 'The six living mermaids.' Beside
the entrance were to be seen pictures of the snakelike bodies of mermaids,
painted with the heads of girls, floating on blue-green ocean waves. The public streamed in in droves, and in a trunk Carl Gottfried Claas Hagenbeck
dragged the shiny silver talers back to Hamburg.—A trunk full of silver talers
in 1848!"[46] What must be recognized is that from the original exhibition of the
seals, through Carl Hagenbeck Sr.'s many exhibits at the annual Christmas fair

(the Hamburger Dom), the fabulous and exotic people shows, and the ventures with circuses and performing animals, Carl Hagenbeck and his father were keenly aware of what the public wanted—of what *entertained* the public— and the company consistently strove to meet those desires.

Indeed, the difference between the elder Hagenbeck's 1858 reconstruction at the Hamburger Dom of the famous attack of the Kreutzberg Circus lion "Prince" on the horses drawing his wagon—which featured a stuffed horse, a stuffed lion, and gobs of red sealing wax—and his son's 1896 reconstruction of a fantastic moment in the history of Arctic exploration is mostly one of scale. In addition to continuing a family tradition, however, Hagenbeck's "Arctic Panorama" of 1896 also took advantage of the current vogue for historical panoramas in German cities. To modern eyes, panoramas, with their lack of action and sound, seem a relatively unimpressive form of popular entertainment. Nevertheless, as Stephan Oettermann suggested in his study of the structures, panoramas completely caught the nineteenth-century public's imagination. At their most basic level, panoramas were semipermanent structures— most often circular buildings—in which particularly dramatic historical scenes such as battlefields or views of famous cities were depicted through a combination of colossal paintings, sculptures, and careful lighting. Usually, the viewer stood on a platform and looked out on a spherical scene that created the illusion of viewing all that existed between oneself and a distant horizon on all sides. The scenes depicted ranged widely but included such moments as an attack against Cameroon "natives" by the German army or the "Arrival of the North-German Lloyd Steamship in the Harbor of New York." The former exhibit "ran" in Berlin from the end of 1885 to the end of 1887, the latter from 1893 to 1895.[47] Perhaps the most famous of the German panoramas, however, was Anton von Werner's "cyclorama" of the Battle of Sedan, which opened in Berlin on September 1, 1883, thirteen years to the day after the momentous scene portrayed actually took place. Present at the opening of the panorama depicting their greatest military triumph were both General Field Marshall Helmuth Karl Bernhard, Count von Moltke, and Kaiser Wilhelm I. The kaiser spent more than an hour and a half surveying the battle scene, standing on a raised platform that rotated at his request so that he could take in the entire view without moving a step. According to von Werner's memoirs, the kaiser was "quite especially pleased . . . by the accuracy of the landscape and the presentation, which in the official accounts of the moment of engagement had not been more closely described, [and] with Moltke [he] followed the course which he had ridden on the 1st and 2nd of September during his visit to the battlefield, remarking to the Field Marshall: 'For the first time I can see here what a long way we went back then.'"⁴⁸

According to a pamphlet describing the exhibit, the cyclorama showed "that moment in the Battle of Sedan during the afternoon of September 1, 1870, between 1:30 and 2:00, when the French Army, namely the VIIth French Corps . . . enveloped by the left flank of the German army and pushed back to the Plateau of Floing-Illy, [was] making its last desperate attempt to smash through the Prussian lines and gain an avenue of retreat."⁴⁹ Dolf Sternberger, in his account of the panorama, argues that the "compulsion for utter illusion becomes all the more obvious" as the pamphlet notes that, "upon stepping out on the platform, the spectator gazes far over the delightful Valley of the Maas; in the depth below him lies the hamlet of Floing, which has been totally occupied by the Germans since 12:30 p.m. In front of the church, the 2nd Comp. of the 82nd Inf. is emerging from the depths. Ibidem: Captain Bödicker (3), commander of the 5th Rifle Battalion, whose 4th company is stationed on reserve in the village."⁵⁰ In describing the effect of the Sedan panorama, Sternberger captures the extraordinary illusionary effect of the best in the medium: "And what a strange blend of real, measurable, spatial distances and their depiction in the pictorial surface; of the imperfections in the human eye and an objective structure in the things it perceives! What a strange blend in the panoramist's expectation of an observer who, as though standing on the generals' promontory, scours the scene with a telescope."⁵¹ The most basic idea of the panoramas was to transport the viewer through time and space to a different world that earlier had been seen only through words in newspapers and books and in flat illustrations. The panoramas, with their unique blend of the real and the unreal, allowed viewers to see and experience the most extreme of adventures in a near-life illusion.

Although directly related to the historical works, Hagenbeck's panoramas were, however, importantly different.⁵² To the historical panorama Hagenbeck added an innovation that was at once both the logical quintessence (the almost real became the real) and ultimately the destruction of the main idea behind the medium: the introduction of live animals. One of the most distinctive features of the panoramas that depicted historical moments was their ability to freeze time. "The Battle of Sedan" caught a specific moment between 1:30 and 2:00

in the afternoon of September 1, 1870, and the visitors to the panorama had the leisure to study that single moment in the greatest detail they desired. If they so wished, they could return another day to study the moment even further; the panorama stopped time forever and allowed the viewer to step outside the moment and to see it from every angle—indeed, as is demonstrated by the kaiser's reaction to "The Battle of Sedan," from angles not possible to those in the real-life drama. In contrast, while it seems that the Hagenbeck panorama of the *Fram* was also built to capture a specific historical moment in the adventures of Fridjof Nansen, the presence of living, moving, and in themselves entertaining animals lent the scene a temporal continuity absent from the traditional historical panoramas. As a result, of course, the story of the *Fram* became incidental, and the focus of the audience quickly turned to the animals themselves. In the end, what made the scene so "extraordinarily fascinating" were the animals and "the appearance of no restricting barrier separating the fierce animals from the visitors."[53] It is not surprising, then, that by the time that Hagenbeck constructed his traveling "Hagenbeck's Animal Paradise—The Zoological Garden of the Future," all historical reference had left the panorama. When future versions of Arctic scenes were constructed—as they were, for example, at the St. Louis World's Fair in 1904—the panorama was essentially the same as that used for the *Fram*, but the ship was gone. Hagenbeck fully realized that the animals themselves and their apparent "wildness" fascinated far more than any dramatic still life.

Thus, by the time Hagenbeck began construction of his famous Animal Park in 1902, and even more so by the time it was completed in 1907, he had already had considerable experience in building panoramas. Over the course of a decade of displaying traveling versions of his exhibits, Hagenbeck had learned what the public wanted, and his park delivered on a monumental scale. His traveling "Zoological Garden of the Future" had been made into a zoological garden of the present, and upon the park's opening on May 1, 1907, enthusiasm seemed boundless.

Not everyone reacted positively to Hagenbeck's new park, however. The directors of zoological gardens throughout Europe were soon questioning the "innovations" in animal keeping supposedly demonstrated by Hagenbeck. To be sure, efforts at acclimatizing various tropical species of mammals and birds so that they could be exhibited outside had been under way in numerous zoological gardens in Europe and in the United States, but in most of these cases the sustained attempts focused on animals that held some potential for profitable domestication in Europe (pheasants, ostriches, and new breeds of cattle, for example). Nevertheless, the directors and architects of the European zoological gardens were in fact sensitive to pleas by the public that the more exotic animals be displayed in larger cages and, whenever possible, outside. As Dr. Kurt Priemel, director of the Frankfurt Zoological Gardens, noted with pride—and also with some frustration—in response to those who found Hagenbeck's park to be something altogether new:

> The zoological gardens have long since, where feasible, broken with the system of cramped cages, and in the last decades especially, enclosures have been constructed which take into consideration the needs of the animals and which show the inhabitants to a degree in a piece of nature. To note a few examples staying only among the German gardens, we should consider the magnificent sea lion exhibits in Cologne and Hamburg, the vast flight cage in Berlin with plants, cliffs, and a waterfall, . . . the constructed mountain cliffs in Berlin and the so-called "Tropical Landscape" for crocodiles in Frankfurt. In these efforts, it is not necessary for the [scientific] methods to step into the background, and therefore we find these ideas being developed as widely as possible in most gardens. Everywhere an obvious scientific aspiration is making itself felt altogether in the gardens, without the general interest suffering.[54]

Joining his voice to the chorus of other directors—including Ludwig Heck, the director of the Berlin Zoological Gardens; Dr. Wunderlich, the director of the gardens in Cologne; Dr. Schäff of the Hamburg gardens; and Dr. Bolan of the Düsseldorf gardens, all of whom had written articles in daily newspapers sharply critical of the enthusiasm for Hagenbeck's park—Priemel acknowledged the public appeal of Hagenbeck's park: "the gardens are described as the 'Seventh Wonder of the World,' as 'The Zoological Garden of the Future'; everything that one sees in Stellingen is supposed to be completely 'new and unique,' the methods used there for the acclimatization and care of exotic animals are supposed to touch on 'totally new principles,' to have been called into being entirely to revolutionize zoo keeping, and 'unsuspected perspectives' are supposed to present themselves."[55] It was precisely this public rather than scientific popularity, however, which drew Priemel's criticism:

> First of all let us be clear that the authors, who can't imagine holding back their judgments from the world, frequently stand quite far from the issues of zoological

gardens, a point betrayed to the expert at so many turns. They see with the eyes of the great masses, exactly for whose visual desires the Stellingen installations were designed. Of all the beautiful and noteworthy accomplishments that Stellingen truly offers, the "great public" sees only the obvious; they pass by the one-of-a-kind collection of walruses, they pay no attention to the famous Hagenbeck antler and horn collection and many other sights, but stand enraptured before the so-called "Grazing Animal Enclosure" and are delighted by the "Lion Grotto" in the background. One sees here on a wide terrain, which because of the great numbers of animals is completely barren of grass, a confusing profusion of representatives of animals from almost every corner of the earth: zebu cattle, bison, camels, llamas, zebras, Shetland ponies, deer, sheep and goats—a piece of "Animal Paradise." The uncritically assessing viewer is, in his enthusiasm, far from able to consider the installation's disadvantages which, to the expert, make it certainly the very least stroke of luck for animal exhibition in this park.[56]

Despite the few notably progressive exhibits in other zoological gardens, however, in most cases the efforts of the traditional zoos to meet the desires of the public for newer kinds of exhibits fell short. Hagenbeck's vision of an *entire* zoological garden in which animals appeared to move unconstrained by imposing bars clearly represented a radical departure, and regardless of how scientifically unsatisfactory the confused jumble of animals may have been or how uneducated the eyes of the "great public" were to the higher purposes of the zoological gardens, this paying public, in fact, stood astounded before the main panorama of the park and was thoroughly won over by the innovative method by which ferocious beasts were displayed. The panoramas at Stellingen did indeed offer the public "a piece of Paradise," and it was a piece that had long been sought.

If Priemel found little to commend at Stellingen, he did nevertheless praise the "Arctic Panorama." He writes: "In a fjord-like ravine, in which massive blocks of ice have been stacked upon each other, animals are shown which belong to the geography and which suit the environs. In the water the most diverse seals and diving birds swim; the background is brought to life effectively through a number of young polar bears, which are separated from the other inhabitants of the installation through a moat which cannot be seen by the public; and on a high plateau one sees a herd of reindeer from the White Sea."[57] Indeed, it seems that to the scientific and zoological garden communities the "Arctic Panorama" was a decidedly more successful exhibit than the "Grazing

Animals Enclosure." Almost any grazing animal from anywhere in the world, aficionados complained, was exhibited in the latter enclosure, and the array confused even the most knowledgeable observer. Additionally, what directors usually considered to be the "best specimens"—the territorial males—could not be placed in the enclosure because of the threat they posed to the other animals. For scientists, the "Arctic Panorama," in contrast, provided the best example of how animals from a specific environment could be displayed together in order to illustrate the relationship between animals and environments.

Often, of course, the scientific interest in particular exhibits led to general interpretations of Hagenbeck's goals which were largely unrelated to Hagenbeck's own intentions in creating the exhibits. Looking back at the park some twenty years after it was built, for example, one of the scientific assistants of the firm, Ludwig Zukowsky, insisted that before all else Carl Hagenbeck wanted to demonstrate how animals live in their natural environments. For Zukowsky, the "Arctic Panorama" provided a perfect example of this goal. He writes: "We see here in the foreground a pool filled with all sorts of seals: seal lions, harbor seals, fur seals, maned seals, and walruses. The elegance of their swimming motions excites quite as much wonder as the plumpness of their motions on lands appears comedic. On the right in the background, separated from the seal pool by a trench which the eye of the viewer cannot see, a group of 8–10 polar bears are located in a cave surrounded by ice blocks; on the left a high plateau is occupied with reindeer, a jumble of fully grown deer with capital antlers, moderately large older animals, and dainty calves looking about stupidly."[58] Zukowsky seems little concerned that Hagenbeck himself appears to have never pointed to a "biological basis" of his exhibits, nor does he seem particularly interested in the origin of the exhibits in paradisiacal panoramas that had little to do with showing how animals actually live in the wild.

Indeed, it is important to realize when looking at the writings of Hagenbeck, Sokolowsky, Zukowsky, and the many others who have analyzed the exhibits in Stellingen that although the exhibits themselves have changed only slightly through the years, the value ascribed to them—their importance, their metaphorical content, their place in the world—has been strikingly flexible. At one moment, for example, an exhibit has been interpreted as presenting the biblical world before the expulsion from the Garden of Eden; at another this same exhibit is described as the Paradise where the lion would lie down with the lamb; at another it is described as having been designed to replicate the natural habitat of the animals by dramatically re-creating a specific biotope; and at yet another, it is considered to have been designed with the interests of the animals themselves foremost in mind. Thus, although after the *Fram* exhibit the Hagenbeck panoramas had ceased to depend on the depiction of specific historical moments, the panoramas were nonetheless filled with narrative content.

If the original metaphor for the Hagenbeck panoramas—already apparent fifteen years before the completion of the park, in the exhibition in Chicago—was of the Edenic Garden or sometimes the Kingdom of God, however, after the park's completion the metaphors changed. From paradises where animals lived side by side in peace, Hagenbeck's park became a sanctuary from a violent world and even a sanctuary from the brutal realities of the evolutionary "fight for survival."[59] Surprisingly quickly, Hagenbeck's park was transformed from a paradise to an ark, a place where animals besieged on all sides in the wild could find refuge in the hands of a congenial old man who became the best friend and perhaps last hope for the animals of the world.[60] As Zukowsky put it in 1929, "In the act of giving his animals, the creatures he loved, a home free of need and misery, Hagenbeck preached that all creatures of the wide, beautiful, roomy Earth had a safe place where they would be secure from the murder and greed of unreasonable and callous people."[61] Pointing to the issue of captivity, Sokolowsky tried to explain the motivations for Hagenbeck's panoramas similarly: "In Carl Hagenbeck the wish grew from his many experiences in caring for and keeping animals—and not least, from his character as a lover of animals—to offer his animals accommodation as appropriate as possible to their nature, a place where they could romp to their hearts' content, and thereby overcome to a certain degree the misery of captivity."[62]

The metaphor of the ark earned the park—and indeed almost all zoos that have since adapted the idea—a profoundly resonant justification for their continued existence in the face of their critics. Hagenbeck's associates, both during his life and after his death in 1913, in fact, have all but suggested that the very future of life on earth rested on the earnest striving of the animal lover Carl Hagenbeck. While noting the laudable efforts of various conservation societies seeking to protect wildlife in the 1920s, Zukowsky, for example, insisted that the only way to prevent the extermination of animals was to teach the masses to love them—and this instruction was provided by both the life of Hagenbeck himself and the "ark" he established to protect them. He writes: "Then comes the great friend of animals, Hagenbeck, and he calls to everyone: come into my beautiful animal park, in my magnificent animal paradise, look at all the diverse creations of God, learn to understand and love them, enjoy them and then go out and protect them across the globe from pursuit and extermination!—And the people come in droves, not simply out of curiosity or the desire to see, but also driven by a longing for nature; they feel that they have lost their connection to Nature. When animals can outdo us in the virtues of courage, faith, and patience, when they can be models for us in their love of their offspring, when they return good deeds with thankfulness and trust, they should not be our enemies, but rather must be our friends!"[63] Indeed, a visit to Stellingen, we are told, not only promoted the protection of the animal kingdom but also restored the essential humanity of men and women in a rapidly changing and "dehumanizing" modern world.

The image of Hagenbeck related by Zukowsky—a completely revisionist reading of the animal dealer—had been firmly entrenched by the 1920s, a time when the company was trying to reestablish itself as the preeminent dealer and exhibitor of exotic animals. This image, however, had its roots, in fact, in the first years of the park when Hagenbeck began to present himself as a congenial Noah figure whose greatest concern was the welfare of the world's animals. Indeed, to a significant extent Hagenbeck's 1908 "autobiography" played a vital part in constructing the various myths about his character; and the park in Stellingen provided the ideal stage for the aging man of the animal trade literally to exhibit his love for animals. In reality, the park became the center of "Hagenbeck's Kingdom"—to borrow the title of Ludwig Zukowsky's history of the firm—the showpiece and public face of a business at the base of which lay the often extremely profitable trade in exotic animals. Though critics such as Kurt Priemel repeatedly attempted to make clear to the public that the park was essentially a showroom for animals that were all for sale, Hagenbeck himself and the rest of the company increasingly presented the park as fundamentally a sanctuary for animals in a hostile world.[64] This was the issue at stake in the photographs discussed earlier of Hagenbeck with his walruses (fig. 21), this was the issue behind the pictures of Hagenbeck in the lions' grotto (fig. 44), this was the message embedded in the memorial sculpture at his grave—a representation of the lioness and "friend" of Hagenbeck, "Triest," lying across his tomb.

If, however, the narrative that the Hagenbecks wished visitors to take away with them had to do with such themes as wonder, exploration, and the exotic,

Fig. 44. Carl Hagenbeck in the lions' den, ca. 1908. *Von Tieren. und Menschen: Erlebnisse und Erfahrungen* (Leipzig: Paul List, 1908). Courtesy of Hagenbeck's Tierpark.

and later, love of animals and conservation, the other narrative of the company, a narrative completely lacking either these "positive" themes or, indeed, their darker counterparts, is saturated with Priemel's pedestrian issues of commerce. In fact, the "real story" of the company can be perhaps best expressed in the many preserved company price lists distributed around the world in the archives of zoological gardens and other institutes. The price lists, sometimes written in Carl Hagenbeck's hand, sometimes typed up by one of his assistants, sometimes printed in little pamphlets or on single sheets, give the values of various animals—the values to Hagenbeck, who had organized their acquisition, and the values to his clients, who wished to own or present them.[65] But what is the value of an animal? How much does a lion, for example, cost in 1900? What has a higher value, an elephant or a tiger? The answers to these questions are anything but simple; they also illuminate as few other facts can some very basic issues behind Carl Hagenbeck's paradise.

First of all, in discussing the value of an animal in the exotic animal market at the end of the nineteenth century, we must recognize that its commercial value was remarkably fluid. Although, for example, an animal might be sold by a company for a given price, during the course of its life in a zoological garden, it might come to be worth a great deal more. All manner of factors could contribute to setting an appropriate price for a particular animal. The age of the animal, its conformity to an expected standard for its species, its sex, its temperament, the presence of certain physical faults (for example, a torn ear, scars, or a limp), its provenance, its potential as breeding stock, and, of course, the current market for the animal—something that could change from month to month—could all play a role in setting its price. Additionally, with the proliferation of species and races, a question about the value of a lion or tiger, for example, becomes, What kind of lion or tiger? Thus, Hagenbeck's price catalogue from November 1901 lists under "Carnivorous Animals":

1 male Barbary Lion, with full black mane, about 5 years old	$1500
1 male Nubian Lion, with black mane, 4 years old	$1000
2 male Nubian Lions, 4 years old, each	$600
1 male Nubian Lion, 2½ years old	$750
2 female Nubian Lions, 2½ years old, very tame, the two	$900
1 female Cape-Lion, imported, 4 years old	$600
1 female Cape-Lion, imported, 3 years old	$600
2 pair Nubian Lions, tame, 16 months old, good for performing purposes, pair	$750
3 male Senegal Lions, imported, very tame, 1 years old, each (good, to form a performing group with the above mentioned 2 pairs of Lions)	$500
1 female Bengal Tiger, imported, 7 years old, good for breeding purposes	$750
1 female Tiger, from North-East-China, about 6 to 7 years old, the right upper fang defected	$700
1 pair Tigers, from Cambodia, 3 to 4 years old, pair	$1500
1 pair Sumatra Tigers, imported, 8 months old, tame as cats, pair	$1000
1 pair imported, very tame Bengal-Tigers, about 1 year old, pair	$1250[66]

Restricting ourselves to just this list, it is clear that a "lion" could cost anywhere from $325 for each of the sixteen-month-old "Nubian" lions to $1,500 for a five-year-old fully maned "Barbary" lion. Issues of the animal's temperament, usefulness, and defects were presented in the price list itself, and further clarifications could be made by cable or other correspondence. In many cases, of course, animals were purchased based on their projected value after several years. In a letter to William Hornaday at the Bronx Zoo in 1910, for example,

Hagenbeck wrote in English: "I just had a letter from one of my Siberian travellers that he has got a fine lot of Siberian animals including some of the large *East-Siberian Tigers*. These are the largest tigers that there are in existence and get the long whool [*sic*] during the winter, but they are animals, which must be kept out in the fresh air winter and summer and dare not be brought into a heated house during the winter. The animals, I expect, will be, on their arrival in the month of May, about 16 to 18 months old, and of course have changed already their teeth. As these animals are very seldom to be got, I could only recommend you to buy a pair of them and would give you a perfect pair, guaranteed from different parents at the price of $2000.—delivered sound and perfect on S/S in Hoboken. You would have to arrange about duty."[67] For Hagenbeck, and for Hornaday, these tigers were not just ordinary tigers but tigers noted for their size and fur and thus could be sold at a higher price. Hornaday, convinced through skins that he had already seen that the tigers would become striking objects at the zoo, purchased a pair of the animals, but in the end he was disenchanted when the young animals arrived in October. While noting that the tigers arrived "clean, well-fed, and in good health," Hornaday nevertheless concludes, "To say that we are disappointed in them is no exaggeration of our feelings. . . . It was our expectation that Siberian tiger cubs would exhibit *some* of the well-known characteristics of tiger-skins that come from that locality."[68]

In any case, although it is clear that many factors contributed to setting a given price for a given animal, certain general features of animal pricing are apparent. Among the carnivorous animals, a particularly splendid adult male lion with an exemplary mane and good temperament, perhaps suitable for training, and female lions with proven success at breeding were the most valuable to dealers and zoos. Tigers followed closely behind the lions and were in turn followed by the jaguars and leopards. Among the last two, the black ones were more highly valued, and, despite their relative rarity in the trade, the various other large cats, including such animals as snow leopards, were rarely valued at the turn of the century at more than a couple of hundred dollars. It should also be noted that crosses between lions and tigers gained a certain curiosity value, and although the animals were sterile, they could easily sell for more than $1,000.[69] For especially fine examples of striped and spotted hyenas, wild dogs, and wolves, dealers could demand between $100 and $200, polar bears could bring as much as $300 or $400, with other bears typically selling for $100 to $200.[70] Among the "hay-eating animals," elephants, rhinoceroses, hippopotamuses, and giraffes constituted the big-ticket items.[71] Young hippos by the turn of the century were going for $3,000 to $4,000, full-grown Indian elephants (seven feet at the shoulder and larger) for around $2,000, young African elephants for as much as double that amount, and giraffes for as much as $4,000.[72] Among the other hay-eaters, prices tended to follow size and rarity. By 1900, American bison delivered through the Bronx Zoo to Hagenbeck were a premium item that could easily cost $1,500, with females costing a good deal more. Similarly, other rare species such as Pere David's deer and the Przewalski horses could sell for more than $1,500. An eland, a particularly difficult animal to obtain in the period, could bring $1,000, a Siberian camel stud $350, zebras (depending on the race and age) from $500 to $800, llamas, alpacas, and guanacos around $200, and the various smaller antelopes and kangaroos $100 to $200; among the deer, fallow deer could be sold for $40 to $50, and larger stags from central Asia with substantial racks could go for $500 to $800.[73]

Among the monkeys and apes, baboons of various types typically sold for $20 to $70 dollars, although larger animals could sell for more than $100, rhesus and other small monkeys for $10 to $50, young chimpanzees for $200 dollars and up, gorillas and orangutans for $500 and up.[74] Finally, among the birds and reptiles, with the exception of a few standout species and individuals, prices were generally under $50. Among the birds, the ostriches, of course, were particularly valuable, especially proven breeders. With a boom in the growth of ostrich farms in the period, adult ostriches could sell for $250 each; emus, cassowaries, rheas, and rarer cranes and storks went for around $100.[75] Parrots and macaws could sell for up to $70, while ducks and swans would go for $5 to $20. Among the reptiles, the larger python snakes could sell for more than $100, with specimens over 20 feet selling for as much as $400.[76]

Of course, a large part of setting the price for an animal turned on the ability of dealers to add value to it. Through appropriate marketing, for example, Hagenbeck could sell animals for a great deal more than might otherwise be expected. Thus, Hagenbeck eventually sold entire groups of performing animals, prepared with certain standard acts, along with cages and wagons for transportation.[77] If Hagenbeck could increase the value of some of his stock by packaging them in novel ways, however, the most typical way for dealer to add value to an animal was to claim it as some sort of rarity. An albino bear, then, however inappropriate for most scientific collections, would always be worth more to a dealer more than a "perfect" representative of its species.[78] Beyond particularly unusual specimens, though, Hagenbeck was perhaps most famous for importing animals that had never been successfully brought to Europe before. Among the animals Hagenbeck was the first to import to Europe, for example, the historian Herman Reichenbach notes the African black rhinoceros (1868), the Sumatran rhinoceros (1872), the Somali wild ass (1882), the gerenuk (1883), the pygmy hippopotamus (1884), the African manatee (1887), the Persian leopard (1897), the Przewalski's horse (1901), the pygmy elephant (1905), the Caucasian wisent (1907), and the elephant seal (1910). Beyond this impressive list, Hagenbeck claimed, of course, to have imported many other species that are now no longer recognized as such. Despite all these efforts, it is important to remember that there were thresholds to the amounts dealers could ask for their animals, no matter how rarely the animal was seen in captivity. Thus, in a case echoing the problems of exhibiting apes, for example, although Hagenbeck tried to sell echidnas to the Bronx Zoo in 1912 for $300 a piece, Hornaday responded that they were "not the most valuable animals in the world to zoological gardens, because in the nature of things they [could] not live very long." Noting that it was "better to invest money in animals that [would] live five years, rather than one year," Hornaday concluded, "The more money we spend for animals, the more we are disposed to put money into animals that are likely to live and give us some return."[79] This said, Hornaday was willing to pay extraordinary sums if he felt that he was getting something truly important or perhaps never before seen. This was the case in the purchase of two of the pygmy hippopotamuses captured by Hans Schomburgk in 1912, for which Hornaday paid the truly amazing price of $12,000, making the two animals certainly among the most expensive animals the Bronx Zoo had purchased up to that time.[80]

I have gone into the issue of animal pricing at such length because, while it is vital to understand the larger environment in which the firm of Carl Hagenbeck operated—a realm in which ideas such as empire, exhibition, and the exotic played compelling roles—it is also necessary to remember the quotidian nature of this business. As can be seen from their surviving correspondence, from day to day and from year to year Carl Hagenbeck and his associates had apparently little time for deep reflection on their activities. Carl Hagenbeck, Josef Menges, Johan Jacobsen, and Christoph Schulz were constantly writing letters, but these letters were almost entirely for business and were directed to trying to secure the best price for this or that item. A number of these people also wrote memoirs, but even in these the focus is on catching animals or organizing people shows and then delivering them to their European and American buyers. Despite, or perhaps because of, the unusual nature of its products, the company's daily concerns necessarily focused on the problems at hand.

As much as the Hagenbeck business was a *business,* it is also important to recognize that Hagenbeck cared about the animals in his park. Indeed, there is ample evidence that Hagenbeck was unusually thoughtful about the animals in his care and that he went to significant lengths to be sure that their needs were being met. With this said, Hagenbeck's Animal Park evolved primarily both to entertain the public and to meet the needs of the various branches of the Hagenbeck enterprise, all of which had at their base the business interests of the firm. Hagenbeck's animal trade, people shows, circus, and Animal Park would not have become operations of such impressive proportions if they had not been profitable; and just as the trade was not primarily organized to protect exotic animals from exploitation, just as the shows of people were not organized to protect indigenous peoples, the park was not built primarily to shelter either the world's exotic people or exotic animals from harm.

Despite the claims made by the in-house historians of the firm—that Hagenbeck's dream was at the very least to give the animals freedom from bars and chains, or, as one of Hagenbeck's animal catchers, Christoph Schulz, put it, to restore to "the captive animals, as much as possible, a piece of their lost homelands"[81]—Hagenbeck's main motivations seem to have been twofold. First, he had learned over the decades before construction of the park began that representations of "authentic" scenes of exotic people and animals, and, more recently, of peace among the beasts, were extremely popular and that an entire park constructed along the principles of the panoramas would likely be a public sensation. Second, Hagenbeck—the man and the company that would soon successfully peddle zoos to individuals and municipalities around the world—wanted to demonstrate that it was possible to acclimate animals completely to northern European environments. This would make possible the construction of relatively inexpensive zoological gardens—zoos without prohibitively expensive buildings and heating systems with budgets that could be used to purchase animals from Hagenbeck.[82] Thus, the underlying impulses behind the construction of the park were ultimately economic. It would be less expensive, especially in the long run, to build "natural" enclosures, and precisely those types of enclosures drew the highest popular interest, thus providing the greatest financial support for the business.

Created primarily for economic reasons, Hagenbeck's "paradise" was at the same time, however, the embodiment of the nineteenth-century public dream of the zoological garden as a place that was able to provide both amusement and education, but an education that would necessarily reinforce a bourgeois view of the world. Visiting Hagenbeck's animal exhibits and people shows, visitors were not confronted with scathing critiques of capitalism, imperialism, or colonial exploitation—this was a idealized world where Europeans could walk among the exhibited animals and people and feel comfortable, secure, and, of course, enlightened. While Hagenbeck's park was clearly a response to nineteenth-century fantasies of the ideal zoological garden, however, it also soon came to represent a vision of what future zoological gardens might be like. Taking its cue from the people shows, the park promoted the idea that a zoo could be a place that sought to re-create the natural world for the viewing public. Re-creating a scene from an Arctic fjord or an African plain, the park attempted to take people to other places of the world, without all the bother and intellectual complication of actually visiting the real place. The message throughout the park, of course, emphasized the benevolent presence of Europeans in a dangerous world while also providing assurance that the animals and peoples of the world were safe as long as they were under the care of Hagenbeck, that is, under the care of Europe. This, it must be emphasized, is the root of almost all modern zoological exhibits. Indeed, albeit for different reasons, Günter Niemeyer, in his 1972 official history of the company, described the patent granted to Carl Hagenbeck for his panoramas as representing "the birth certificate of all modern zoo construction." There seems to be almost universal agreement with Niemeyer's judgement.[83] In a quite traditional article on the architecture in zoological gardens of the 1970s and 1980s, for example, Rosl Kirchshofer restated the usual argument about Hagenbeck's contributions quite succinctly: "With his zoo in Hamburg-Stellingen, which was opened in 1907, [Hagenbeck] not only released the 'beasts of prey' from their cage-like dungeons, but by providing a view of animals moving freely in a designed environment, he also opened up the way for a better understanding of them on the part of the visitors."[84] Heine Hediger, perhaps the foremost theoretician of the zoological garden in the twentieth century, suggested a similar heritage for the modern zoo. In his *Man and Animal in the Zoo,* from the late 1960s, he records, for example, that "even during the heyday of the iron bar method of fencing, a development along completely different lines was introduced at the beginning of the century; this was the so-called barless *open enclosure,* based on an idea conceived by Hagenbeck."[85] Although Hediger notes that Hagenbeck's concepts were "laughed at and ridiculed to start with," he traces a direct line from Hagenbeck's innovations to modern ideas about animal enclosure.[86]

Of course, by emphasizing the benevolence of Hagenbeck and the colonial powers, the promoters of Hagenbeck's park were forced to fabricate a heroic history for both Carl Hagenbeck and his various business concerns, including the people shows and the animal trade. Thoroughly convinced of the need to protect the world's animals against their wanton destruction by thoughtless people, for example, Ludwig Zukowsky cites cases of human destructiveness:

> How many fantastic animal types have already been destroyed through the centuries-long campaign of extermination, and how many stand already on the list of endangered species. Many animals are still regarded only as items of trade; their mass-slaughterers feel sorry certainly like that Roman Emperor, that the whole living world had not a single head. Were Mother Nature not so extravagantly fruitful, and were not her children so carefully divided up around the world, their end would already have come before the current human generation. How many fur seals, walruses, and elephants have already been butchered! In 1886 on the northern coast of Nordostland in the North Sea, a walrus herd of 370 individuals was killed by the crew of a ship. Mother Nature was 370 children poorer, the ship 370 skins richer.[87]

After chronicling that once 700,000 of 800,000 seal skins were burned to protect their market price, Zukowsky concludes, "These are tragedies in the animal world and documents of the time of shame. Someday the grandchildren will accuse their fathers and call to them in their graves: 'You have squandered our inheritance, destroyed the beautiful world and left us nothing but rubble!'"[88] This all-out condemnation of the greed of dealers in wildlife products who continued to treat exotic animals "still only as articles of trade" is followed by passages quoted earlier in which Zukowsky claims that it was Hagenbeck who responded to the slaughter by asking the public to come to his park and learn to love the animals.

What is truly extraordinary about Zukowsky's account is that he appears to have discovered the story of the mass slaughter of the 370 walruses from the autobiography of Hagenbeck himself. In his discussion of walruses, Hagenbeck relates the anecdote to illustrate his point that during the breeding season male and female walruses converge in large colonies. Hagenbeck writes:

> The sexes of the walrus keep separate from one another, the females and young live for themselves, as do the males, and only in the breeding season in September and October do males and females come together on land. At such a time, Captain Hansen ran into a herd of 370 individuals in 1886 on the north coast of Nordostland. They were all killed by the crews of five ships. Otherwise, the females live, according to the reports of the men, on the north coast of Spitzbergen, at about eighty-one degrees north latitude; the males, on the other hand, stay in the *Storfjord* between Nordostland and King Karlsland.
>
> The largest bull, however, that our catchers came across on their last voyage and fortunately killed, had a weight of approximately 3000 kilograms. The skin alone weighed 500 kilograms.
>
> The young animals were caught near Cape Flora, but the most fertile hunting ground these days is the north coast of Siberia.[89]

At no point in Hagenbeck's account is there any suggestion that he finds the "killing"—what Zukowsky called "slaughter"—of the herd of walruses either "tragic" or "shameful." The story is told because it establishes the potential size of walrus herds, and it is clear that Hagenbeck's sympathies are decidedly on the side of the hunters. Explicitly pleased that the largest bull in the herd had "fortunately" been killed because of the value of its hide and blubber, for example, Hagenbeck seems to be far from Zukowsky's revisionist image of the kind Noah.

In a similar act of generous interpretation, Sokolowsky relates what appears to be an apocryphal story of a great hunter who had lived and hunted in Africa for many years. One day the man visited the park and told Hagenbeck a number of his hunting adventures. According to Sokolowsky, when the man had finished with his tale, "Herr Hagenbeck asked him, quite surprised, the following question: 'And you just come right up to me to tell me how many animals you have shot? How then am I supposed to catch and sell animals, if you shoot them all to death!'"[90] For Sokolowsky, Hagenbeck's remark signified his disgust with people traveling to Africa merely to hunt and then to decorate their walls with trophies of their slaughter.[91] A more simple interpretation, however, might suggest that Hagenbeck was at least as concerned about the difficulty of pursuing his business in the face of such hunts as he was about the extinction of animals around the globe.

There is, in the end, a striking disjuncture between the actual nature of Hagenbeck's park and the ideas that have been placed into it. Priemel tried to clarify the distinctive quality of Hagenbeck's park in 1909: "Carl Hagenbeck's Animal Park is a large-scale *display* and *trading enterprise* with the virtue of possessing in some of its departments even some scientific value; the *zoological gardens* on the other hand are in the first instance *scientific, educational institutes* which must accept restaurants, concerts, 'illuminations' and other such activities only as trimmings to gain the necessary operating finances, to a certain extent as necessary evils, a fact which all too often, especially in Frankfurt, is unrecognized."[92] Hagenbeck's park was simply neither a paradise for animals nor an ark for their protection. In most cases, it was a holding station for animals brought from all over the world which were waiting to be shipped to buyers. During the time the animals were in Stellingen, they were part of a large-scale, almost exclusively entertainment-oriented medium designed to meet popular fantasies about animal life. While Priemel and other directors of zoological gardens bemoaned the need to meet the entertainment desires of the paying public, Hagenbeck's park was specifically designed to meet those desires as a type of a proto-Disneyland where horses, ostriches, and zebras pulled carts around and camel, elephant, and tortoise rides were available to all who paid for the pleasure.

Of course, when people came to the park, they rarely understood that the animals they saw were essentially in transit. But the largely hidden quality of Hagenbeck's park/ark as a depot should not be underestimated. On his first day as a scientific assistant at Hagenbeck's park in 1913, for example, Alexander Sokolowsky recalled in a brief memoir for the Association of German Zoological Gardens that he was sent to inventory a newly arrived ship from Australia. On board he found 40 kangaroos, 26 emus, 600 finches, and some 300 reptiles of various species. Meanwhile, Christoph Schulz had just returned from German East Africa with a shipment of 36 white-bearded gnus, 25 Böhms zebras, 3 black rhinos, 5 giraffes, various species of antelopes, and nearly 300 monkeys, great cats, and smaller animals. Soon Hagenbeck's man in Siberia, Osip Neschiwow, arrived with a large collection. Sokolowsky writes, "The number of . . . rare smaller animals went far into the hundreds, while species of larger animals were on hand in the dozens."[93] Included in the shipment were 200 Chukar hens, hundreds of ducks, and 200 ring-neck pheasants. On another occasion in 1914, Neschiwow delivered, "besides the usual mammals and birds, . . . no fewer than 8,000 top quality clean Horsfield's tortoises packed in sacks."[94] With Neschiwow's departure in 1913 came the arrival of another Hagenbeck agent from East Africa with 250 baboons. A week later, 21 elephants and 12 Bengal tigers arrived. After several years of being closed because of difficulties following World War I, the Hagenbeck business picked up again in the 1920s, collecting and shipping, for example, some 8,000 Rhesus monkeys to the Rockefeller Institute in the United States to be used in the study of yellow fever. In 1929, Sokolowsky recalled, there were 250 penguins of five species at the park, and during his years at Stellingen, he concluded, he must have seen some 2,000 penguins of eleven different species. These are the numbers behind the panoramas of paradise at Hagenbeck's park.

In a world in which zoological gardens exist, animal dealers and traders must also exist. As Karl Max Schneider of the gardens in Leipzig wrote in 1954: "We directors of zoological gardens know quite well that a strong animal trade is necessary. We need those kind of middlemen. The regular acquisition of animals from foreign countries is impossible for a garden or even a group of them."[95] In part, it was the existence of these "middlemen" which made it possible for most zoological gardens over the course of the twentieth century to fashion themselves as conservators and protectors of animals from exploitation. With that said, it was precisely one of those middlemen—Carl Hagenbeck—who constructed the essential and lasting basis for zoological gardens in the twentieth century. In his park, we are told that animals live in complete freedom while enjoying the beneficence of a kindly lord.

Looking out over Hagenbeck's Animal Park today, one can be struck by how little it has changed over the course of almost a hundred years. Expanded and refined in the late 1920s and throughout the 1930s, the park was almost completely destroyed in the Allied bombing raids in 1943. The park was rebuilt, however, and in keeping with Carl Hagenbeck's remarkable vision of the zoological garden; in the park today, most of the original panoramas stand with little apparent modification. To be sure, much has changed. Most fundamentally, the company—remarkably still held privately by Carl Hagenbeck's descendants—is no longer an animal dealership. The people shows, moreover, are gone, although echoes of them can be seen in occasional cultural spectacles.

Most important, however, as unchanged as the physical structure of the park may be, the philosophical basis of the Hagenbeck enterprise—already changing in the years after its completion in 1907—bears little resemblance to Carl Hagenbeck's interests. Nevertheless, the illusion of freedom created by Carl Hagenbeck—with the complementary illusion that the exotic animals of the world have at least refuges in cities where they will be cared for even if their kind dies out in the wild—continues.

"My favorite author? Oh, Darwin, of course."
—The gorilla Miss Crowther "interviewed" for the London *Star*
in 1905

Over three years ago a party of Hagenbeck's animal hunters, headed
by Schultz, were trekking through one of the wildest sections of the
Cameroons when they encountered a huge male gorilla with a cub.
—On the catching of the gorilla John Daniel, *New York Times,*
April 3, 1921

Surrounded by strangers and visitors, seeing no one that understood
him, or whom he liked, the inevitable happened: John became fatally
lonesome, lost appetite, lost courage, and finally fell ill. . . . John
Gorilla died in the last week of April, 1921, a week before his beloved
friend arrived.
—William T. Hornaday on the death of John Daniel (John Gorilla),
Mentor, November 1921

The story of Red Peter in Kafka's "A Report to an Academy" is central to this book. Like most works in the animal
story tradition, from Aesop's fables and Apuleius's *Golden
Ass* to more modern stories by Swift and Orwell, "A Report
to an Academy" is both about humans and about animals,
or rather about how people imagine animals. Most scholars
who have examined animal stories will point out, of course,
that the animals in these works are really only incidentally
animals; the stories, in other words, are in essence about
people. Nevertheless, I think it is important to bear in mind
that, while it is true that the medieval world described in
the French stories of Reynard the Fox is very human, there
is also an element running through the Reynard stories
which tells us something about what people thought about
foxes in the twelfth and thirteenth centuries. When I first

read the story of Red Peter, it seemed clear enough that, although the story was centrally about problems in human society, it also was about the problems faced by an ape obtained by one of Hagenbeck's catchers and brought to live among civilized humans. Just as we should remember that there were a great many real historical figures behind Kafka's "The Hunger Artist," Red Peter is in fact based, at least to some extent, on a whole line of real and quite famous apes who were—to use the title of one of the classic works of twentieth-century primatology from 1925—"almost human."[1]

This point is made even clearer if we recall virtually any of the many accounts of real nonhuman primates living in Europe or the United States in the forty or so years around the end of the nineteenth century and the beginning of the twentieth. All these stories—like that of Red Peter—work simultaneously on both literal and metaphorical levels. Consider the chimpanzee "Consul," shown dressed in an evening suit before a tea service while holding a knife and fork on his ringed fingers (fig. 45). Consider the gorilla "Miss Crowther," "interviewed" for the *Star* in London about her impressions of England during

Fig. 45. Consul, the civilized ape, ca. 1900. From Frank C. Bostock, *The Training of Wild Animals* (New York: Century, 1903).

Fig. 46. Primate tea party, ca. 1908. *Von Tieren und Menschen: Erlebnisse und Erfahrungen* (Leipzig: Paul List, 1908). Courtesy of Hagenbeck's Tierpark.

her brief residence in the fall of 1905 at the London Zoo. Consider the famous Ringling gorilla of 1921, John Daniel (also known as John Gorilla), who, like Red Peter, was caught by one of Hagenbeck's agents and was raised to live as a human before being sold to the circus. Consider the gorilla John Daniel II, who was greeted at the docks in New York City in April 1924 by a welcoming committee headed by one of the most famous "missing links," the freak William Henry Johnson.[2] Consider the almost ubiquitous tea parties for primates (fig. 46) staged in major zoological gardens up through the mid-twentieth century. Consider the many images from the period of very young apes in baby carriages and somewhat older apes drinking wine and schnapps (figs. 47 and 48).[3] In all these instances, the animals were trained to be human-apes, to be both animal and not-animal, human and not-human—and it was precisely their in-between status (like that of the character Red Peter) which allowed them to be used to critique both human and animal lives.

What seems to distinguish Red Peter from ordinary apes and other animals, of course, is that Red Peter can speak, and indeed he speaks with particular eloquence. Adopting a vocabulary and style suited to his academic audience, Red Peter addresses the members as a fellow colleague with unique insights into questions of evolution. Red Peter begins:

> Honored Members of the Academy!
> You have done me the honor of inviting me to give your Academy an account of the life I formerly led as an ape.

I regret that I cannot comply with your request to the extent you desire. It is now nearly five years since I was an ape, a short space of time, perhaps, according to the calendar, but an infinitely long time to gallop through at full speed, as I have done, more or less accompanied by excellent mentors, good advice, applause, and orchestral music, and yet essentially alone, since all my escorters, to keep the image, kept well off the course.[4]

Indeed, it was precisely when Red Peter spoke his first word, after having finally overcome, at least to some degree, his revulsion with schnapps, that he "broke into the human community."[5] Human language was the great achievement of Red Peter; the rest of his remarkable career followed by necessity. To note that Red Peter was different from other animals because he could speak, however, does not get to the heart of Kafka's story. Even if Red Peter could speak, his nonspeaking captive comrades could nevertheless be conspicu-

Fig. 47. Orangutan in a baby carriage, Fockelmann's animal business, Hamburg, ca. 1908. From Alexander Sokolowsky, *Beobachtungen über die Psyche der Menschenaffen* (Frankfurt: Neuer Frankfurter Verlag, 1908).

Fig. 48. Moritz at Hagenbeck's Animal Park, ca. 1908. *Von Tieren und Menschen: Erlebnisse und Erfahrungen* (Leipzig: Paul List, 1908). Courtesy of Hagenbeck's Tierpark.

ously—and often disturbingly—eloquent. Indeed, Kafka's tale makes sense precisely because human audiences at zoological gardens, circuses, and variety stages had been "hearing" animals "speak" for decades. Of course, some of the animals were more rhetorically accomplished than others. A bull from India was unlikely to "say" much, but an elephant, as we learned from the Munnecke stories, could perhaps have many stories to tell. Breugel's monkeys chained in a window comment on captivity as much as Rilke's panther: "The silent shutter of his eye sometimes / slides open to admit some thing outside; / an image runs through each expectant limb / and penetrates his heart, and dies."[6]

I am not arguing that the panther that Rilke saw at the Jardin des Plantes in 1907 actually expressed what Rilke "heard." Rather, I am arguing that Rilke, reflecting on the panther turning its looped path in its small cage, wondered what it must be like to be a panther in a cage. This is the very sort of question which was posed by Kafka when he presumably asked himself what it must be like to be an ape in a cage on a ship bound for Europe. So much of Hagenbeck's exhibitions, both animal and human, appears to have been directed to answering, or at least redirecting, those questions; Hagenbeck's exhibits with their contented people and free animals answered the public's concerns over captivity with a gentle smile. Much of the purpose of this book has been to try to raise the questions once again.

Permeating these questions are the peculiar tensions between Hagenbeck's various businesses, most specifically between his exhibitions of people and his exhibitions of animals. Not surprisingly, most people hearing about Hagenbeck are taken aback by descriptions of the people shows. They ask incredulously, "You mean, this guy actually exhibited people in a zoo like they were animals?" In response, I say emphatically that I do not believe Hagenbeck exhibited people as animals; he very clearly exhibited them as people. As I have argued, though, in order for these exhibits to succeed, two elements were essential. First, audiences had to be convinced that what they saw before them was actually what the company claimed—that the exhibited people were "authentic." Second, they had to be convinced that the exhibited people were fundamentally different from and inferior to themselves—the "natives" could be admired for their naturalness or beauty, but they nevertheless had to be seen as inferior to Europeans in the most important cultural accomplishments. The demise of the shows, in fact, appears to have been the result of the refusal of exhibited peoples to comply with the basic dramatic premise of the displays. Rather than be incomprehensible and uncomprehending savages, the people in the shows learned German, learned to ask for tips, enjoyed visiting taverns, purchased European clothes, participated in sideline commercial activities involving sexual and other services, and otherwise confounded an audience who came to the zoo to see savages behaving as savages were expected to behave. Indeed, precisely because Hagenbeck's exhibits purported to avoid all theatricality, to present only what was true to life, they were especially vulnerable to the exhibited people being, in point of fact, true to life. The exhibits came to an end because, in the final analysis, it was impossible to organize shows of "savages" when one couldn't find any. The "savages" themselves ceased to exist when the indigenous people in the exhibits refused to play their roles.

Carl Hagenbeck published his memoirs less than a hundred years ago, and if it is clear to us that exhibiting people presents a different set of problems than exhibiting animals, this was probably clear to both Hagenbeck and his contemporaries as well. At the same time and not surprisingly, Hagenbeck and his overwhelmingly supportive audience tended to emphasize the "positive" aspects of his people exhibits. Alexander Sokolowsky, for example, argued: "Carl Hagenbeck not only accomplished a great deal as an animal friend and animal importer for the education of the people as well as the advancement of science, but through his extensive exhibitions of people he also played at least as great a role in the broadening of our knowledge and thereby our cultural understanding. To all of us, he was a teacher whose life-long work—the advancing of knowledge—will continue to have an effect. He awoke the understanding for the more primitive peoples in the life of man, made possible a comparison between the diverse exhibited peoples, and eliminated many prejudices."[7] Far from being an exploiter of indigenous people from around the world, Hagenbeck, we are to understand, loved all people and wanted more than anything else to educate his public, facilitate "understanding," and dispel "prejudices." As I have suggested throughout this book, however, Hagenbeck and his associates faced considerable predicaments in putting a positive face on a company that was based on the capturing, trading, and exhibiting of animals and people. Indeed, that the company has over the past century gone to great lengths to convince its public that the congenial old Hagenbeck should be seen as a modern Noah and that the Animal Park in Stellingen should be understood as an amalgamated Ark in Paradise speaks clearly to the ironies inherent in Hagenbeck's diverse enterprises.

These ironies are thoroughly embedded in the company's history. Consider, for example, the photograph of a young gorilla and two youths from Cameroon (fig. 49) which appeared at the end of Hagenbeck's memoir. According to Hagenbeck, a lieutenant in the German colonial army in Cameroon brought the gorilla to Germany in June 1908 in the company of the two young boys. Hagenbeck writes that the officer had

> hoped to be able to keep this rare animal alive for a long time. Over in Kamerun he had kept it for more than a year, during which time it had enjoyed unbroken health and become a general pet of the station. He hoped to be able to overcome the difficulty of lack of society by providing the two negroes as constant associates for the animal. When the ape first arrived at my animal park he was much weakened with his long sea voyage and took little interest in anything that was going on round

about, but he soon picked up, and after a time would sit and walk about on the lawn in company with his two play-fellows, apparently in the best of health and spirits. He had a strong predilection for the petals of roses, and would consume large quantities of them. When he had to be taken from one place to another one of the negroes used to carry him on his back, presenting a very droll appearance.[8]

Responding to the popular assessment of the time (articulated at the beginning of this study) that, before all other causes, gorillas in captivity died from depression and loneliness, the officer secured two young boys to accompany his gorilla to Europe and to live with the animal until, presumably, it either died or was sold. The arrangement must have seemed perfectly sensible to officer, perfectly sensible to Hagenbeck, and perhaps even perfectly sensible to the boys themselves. All this sensibility aside, however, the photograph retains a deeply unsettling quality that is only amplified by its caption in the German edition: "Prophete rechts, Prophete links, das Weltkind in der Mitten" (Prophets to the right, prophets to the left, the worldling in the middle).[9] Stemming from a small humorous poem by Goethe commemorating a dinner in 1774 at which he sat between the physiognomist Johann Lavater and the educational reformer Johann Bernhard Basedow, the caption, it seems, is meant to add a certain lev-

Fig. 49. "Prophets to the Right, Prophets to the Left, the Worldling in the Middle," 1908. *Von Tieren und Menschen: Erlebnisse und Erfahrungen* (Leipzig: Paul List, 1908). Courtesy of Hagenbeck's Tierpark.

Savages and Beasts

ity to a picture with little obvious humor. According to accounts of the dinner, while Lavater and Basedow carried on at length with their various remarkable ideas, the young Goethe sat quietly and devoted himself to the food—while two thinkers concerned themselves with matters of the mind, a sensualist attended to more immediate concerns.[10]

What is it about this picture that is so disconcerting? I have shown it to a number of people who have been part of my concluding efforts in this book, and their responses have been significantly varied. An art historian said it reminded him of a modern Golgotha (that is, two thieves flanking one savior); a scholar of Victorian literature felt that the picture was of shame, including the shame of the viewer; a specialist in American ethnic literature brought up the issue of race; a studio artist suggested that the picture seemed to tell a story about the relative value of the three figures, a story, that is, of prophets and profits.[11] My central reaction, which should not come as much of a surprise at this point, is that the picture is about the problems of showing captivity. The photograph is disturbing, I believe, because in its wearied quiet it is deeply receptive to narrative. Seeing a photograph like this, we begin to imagine stories to explain it, and it is those stories that Hagenbeck's exhibits—whether of animals or people—sought to control. Without the caption, we are let loose to interpret the photograph according to our own sensitivities. With the caption—and its suggestion of a parallel between a young and worldly Goethe and a young and worldly gorilla—the viewer is asked to understand the photograph as somehow amusing. To return to the idea of eloquence, the inherent eloquence in the expressions of the two young boys and the young gorilla is managed and constrained by an amusing caption and Hagenbeck's story of their "droll" visit to the Animal Park.

Indeed, "managing eloquence"—attempting to redirect the audience from seeing and imagining an animal's fate in captivity—is perhaps the fundamental feature of Hagenbeck's park. The illusion of freedom so carefully created in Hagenbeck's innovative exhibits and so carefully maintained in such modern zoological gardens as San Diego's Wild Animal Park and Disney's Animal Kingdom controls the expressiveness of the animals. In a photograph of an animal sitting hunched in a cage (fig. 50) or pacing before a staring audience munching on popcorn, hotdogs, and cotton candy, for example, we can immediately read issues of captivity. In our new zoos, on the other hand, with their carefully deployed plants and illusions of freedom which trace back to Hagenbeck's

Fig. 50. Jacob at Hagenbeck's Animal Park, ca. 1908. *Von Tieren und Menschen: Erlebnisse und Erfahrungen* (Leipzig: Paul List, 1908). Courtesy of Hagenbeck's Tierpark.

park, the person poking a rhino with a stick to get it to move is shunned. Now we see animals moving quietly in the woods, gathering at a water hole, and lounging in satisfaction in the afternoon sun on a kopje. These idyllic settings mask the fundamental nature of an animal's captivity, and if the people in the people shows resisted the narrativizing strategies marshaled by the company, the animals in our new zoo exhibits, surrounded by plants and fabricated trees, face a much more difficult time finding a voice with which to query their audience.

At the beginning of this book, I noted that around the end of the nineteenth century directors and other proponents of zoological gardens began to formulate the raisons d'être of zoological gardens of today. The ideas, revitalized again and again during the past hundred years, have had remarkable staying power. The fundamental purpose behind the arguments was to make clear to the public that modern zoological gardens were fundamentally different from their historical predecessors. The bourgeois zoological gardens of the second half of the nineteenth century, authors argued, represented something entirely different from the old princely menageries. Whereas the aristocratic collections were little more than exclusive pleasure gardens, the zoological gardens, often built through public subscription, were institutions founded for three basic reasons: to advance science, to promote public education, and to provide people with a refuge from the pressures of urban environments.

Then came the "Hagenbeck revolution" just about the time that a new claim about the purpose of the modern zoological garden began to be elaborated. Facing issues such as the disappearance of the American bison and concerns over the commercial slaughter of animals in Africa, zoo advocates began to argue that the gardens might serve as sanctuaries not simply for a weary public but for a threatened animal world. So what precisely was the Hagenbeck revolution—the revolution to which our contemporary zoos consistently trace their origin? The answer that one generally hears is that Carl Hagenbeck invented a way of exhibiting animals by exploiting moats and other techniques that did away with both elaborate buildings and barred cages. This, I believe, is a small point. Probably every major zoo director at the end of the nineteenth century was aware that the bars on the cages represented a problem for visitors, and a good many zoos had been experimenting with different kinds of exhibits as a result. Indeed, in the final analysis, and despite the consensus, Hagenbeck's revolution was not really the moated structures he created. Hagenbeck's revolution was precisely the narratives of freedom and happiness he developed at his zoo to go along with the newer exhibits. Before Hagenbeck, zoological gardens often struggled to convince the public that it was not so bad to be an animal at the zoo. Beginning with Hagenbeck, the gardens began finally, and more or less successfully, to renarrate the captive lives of animals. Ever since Hagenbeck, animals have not been collected merely for reasons of science or education, or even really for recreation—animals have been put in zoos increasingly because they are nice, healthy, safe places to be and because the animals, we are told, might be better off there than in the real "wild."

This is not the place to explore at length the history of the zoo in the twentieth century. That history is complex and contains a remarkable array of unexpected twists and turns. There are, for example, moments in which developments in modern art and architecture seem to play a larger role in designing animal exhibits than anything else. Indeed, far from the thoroughgoing representationalism of Hagenbeck's exhibits, a number of projects at the London Zoo—including the Bauhaus/Internationalist–inspired Penguin Pool of 1934 and Gorilla House of 1932–33, both by Tecton—seemed to be taking animal exhibits more in the direction of a Stanley Kubrick space station than to recreations of life in the wild. Later, the London Zoo's Elephant and Rhino Pavilion of 1965 seemed to attempt to create an idea or abstraction of "nature" without resembling anything that could be actually be seen *in* nature. From the outside, people stood before a low wall with a chest-high bar and looked across a small moat at the conglomerate form of the central pavilion—an imposing, light-earth-toned, rough concrete building that has been described as "zoomorphic New Brutalism, marvelously expressive of its inhabitants."[12] At the top of the building, huge copper-sheathed projections rose to collect light for the interior of the building. But instead of nineteenth-century skylights lighting up the public spaces of the interior, in this abstract building the light fell only on the animals in their interior enclosures—the public now stood in the dark. The people moved quietly in an increasingly nonhuman world. This is not a scene in which human sociability or the pet-making practices of the nineteenth-century zoo could find a place. Rather, this was an abstracted home for wild animals in which humans were admitted only under restricted conditions.

Despite the various exhibit experiments that have been constructed in zoos over the almost hundred years since Hagenbeck's park opened—despite the antiseptic "bathroom-style" primate houses of the 1950s and proposed plans by the architectural firm of Cambridge Seven Associates for a vertical zoo in downtown Boston—the twentieth century ended just about where Hagenbeck started it: the newest zoo designs for "immersion exhibits" have returned to his ideas of re-creating natural environments for the animals.[13] The dream of these exhibits is captured well by Vicki Croke: "At the San Diego Zoo, a visitor may find herself alone on the steep and shady path that leads down to the Tiger River exhibit. The vegetation on either side is dense and lush. The visitor is 'immersed' in the exhibit, transported into a jungle in Asia. But this jungle is inhabited by tigers, and when a low growl is heard just beyond the next bend, the temptation is to run."[14] This is precisely the sort of language which people used to describe Hagenbeck's park when it opened in 1907.

To be sure, with every advance in technology, immersion exhibits become more elaborate and more spectacular. In the new, more perfect worlds of the immersion exhibit, a better "nature" is created for animals: food is plentiful, more and more interesting, and a mental challenge to find; parasites are carefully managed; sicknesses are combated with the full range of modern medical technologies; climate is thoroughly regulated by advanced computer systems; human visitors are literally engulfed in the exhibit and yet know they shouldn't intrude on the animals' activities; sounds that seem somehow fuller and clearer than the actual sounds one might hear near forests, seashores, and mountain escarpments are piped in through camouflaged speakers; and successful propagation is the clear measure of happiness and health. These exhibits seem like they must be wonderful places for animals.

It is crucial that we remind ourselves, however, that from the very beginning of Hagenbeck's experiments with what we now call immersion exhibits, it has been clear that this type of exhibit was designed for the pleasure of the public, not the animals. When people came to Hagenbeck's park and saw the animals living in apparent freedom, they were ecstatic. But Hagenbeck's park was not about restoring animals to their natural environments. It was better than that. As should be clear from a photograph of the park's main panorama (fig. 43), the immersion exhibit was never really intended to trick people into believing they had stepped into a natural scene. Just as those people visiting the huge "Congo Gorilla Forest" at the Bronx Zoo know that they have not been miraculously transported to the west coast of Africa, Hagenbeck's goal was not simply accurate simulation. His goal—and that of all designers of immersion exhibits—was to convince people to suspend their disbelief long enough to accept what they saw before them as an alternative but believable scene. The goal of the immersion exhibit was and is to create a convincing verisimilitude. But to be convincing, it seems necessary for the immersion exhibit to actually outdo nature. Compressed into small spaces, the better nature of the zoo makes real nature seem dull in comparison. The nature of the zoo suggests that there should be an animal—or better yet many animals—in every scene and that all one has to do is look hard enough to find them. But is all this good for the animals? Or more broadly, does the attention lavished on a particular gorilla or pair of young pandas in Atlanta yield improved chances for their species?

The answer to this seemingly simple question is not easy. Consider, for example, the case of Keiko the Killer Whale, who starred in the movie *Free Willy* in 1993. Through the broad sentimentalization of Willy/Keiko, millions of dollars continue to be raised and spent to return the whale to the wild. Keiko is a first-rate animal star, and images from the Keiko-cam (one of the first of the now almost ubiquitous zoo-cams continually posting pictures of zoo celebrities to the Internet) were downloaded by the thousands while the whale was convalescing at the Oregon Coast Aquarium. It seems reasonable to suggest that the story of Keiko has indeed made some people more aware of the difficult lives of large marine mammals in zoological gardens and aquariums. What is also completely clear, however, is that the major commercial aquariums have not only endured the criticism surrounding the life of Keiko but have actually become more eager to have whales—and other particularly charismatic creatures—in their collections because people want to "see Willy."

Zoos exist because people find them interesting, relaxing, fun, or educational places to go. More and more realistic exhibits at zoos exist not because they tend to lessen the amount of stereotyped behavior seen in the animals (which they often do) or even because animals often find contact with real soil and plants interesting and enjoyable. They exist because people have come to dislike looking at animals behind bars and in small glassed-in rooms and prefer exhibits in which animals appear to be living in nature. But, as the history of Hagenbeck makes clear, the new immersion exhibits are not "nature." They are not even replicas of "nature." They are fantasies now reinforced by nature television, in which at every turn the camera seems unbelievably ready—and the light implausibly perfect—to catch the most unimaginable shot.

In looking at the history of the Hagenbecks—whether studying the animal trade, the people exhibitions, the performances of trained animals, or the animal park—historians should try to understand, before all else, the motivations behind the various endeavors. Those motivations are rarely simple. It seems, for example, that the two most basic impressions immediately drawn of Carl Hagenbeck are that he was a dealer in animals and that he was a lover of animals. Both of these perspectives tell part of the truth, but neither adequately explains the complicated interests and activities of Carl Hagenbeck and his firm. Ever since the late nineteenth century, Hagenbeck and his followers have chosen to concentrate his image into only one readily accessible concept—Hagenbeck as a friend of animals and people. From a man who had made a considerable fortune organizing the capture, purchase, and sale of tens of thousands of animals and people, the company developed a story of a man who was the ultimate friend of living things, a man whose goal in life was to spread appreciation and understanding among the people who came to his sanctuary.

Despite these assertions, the one salient feature linking all Hagenbeck's ventures involves their entrepreneurial aims. Each branch of the firm became what it was because of its profitability. Carl Hagenbeck would not have continued to organize people shows—especially after the disastrous deaths of the troupes from Terra del Fuego and Labrador—if the exhibits had not been popular and profitable and had not provided a successful alternative to an animal business that was experiencing a slump. Similarly, Carl Hagenbeck did not become the leading animal dealer in the world simply because he loved animals. He became so because of his extraordinary business sense, the accelerating growth of a market for the animals, his location at one of the hubs of international exchange, his carefully nurtured contacts in the zoological garden and circus world, his desire to bypass the traditional lines of bringing animals to Europe by sending his own catchers into the field, and his plain hard work.

Confronted with the contrasts presented by the stories surrounding a firm such as that of Carl Hagenbeck, we are left with some serious questions. Pictures held in the Hagenbeck archive, for example, show giant tortoises from the Galápagos Islands. In one photograph the tortoises are stacked high in the storage area of a boat used to transport products from ship to shore, and in another we see finely dressed children riding the creatures before the Hagenbeck villa at the park. While it is completely one sided when looking at such pictures to conclude that Hagenbeck was a ruthless exploiter of animal life whose wealth rested on the corpses of animals spread across the globe, it is equally problematic to conclude that Hagenbeck was a modern Noah in whose park the animals of the world were forever safe. What we can and should say is that in his day few could seriously criticize Hagenbeck for pursuing his entrepreneurial visions. During his final years, in fact, Hagenbeck was far from a villain; indeed, he achieved heroic status. He was the man, for example, to whom the government turned to organize the acquisition of two thousand dromedaries—with saddles and caretakers—for delivery to the German troops battling the Herero resistance in German Southwest Africa.[15]

Regardless of the stand one takes on Carl Hagenbeck, historians must accept the responsibility of questioning the ways in which the life of a man, his actions, and his motivations have been reinterpreted by those who have followed him to create and perpetuate misleading myths. Through affection, admiration, and loyalty, Hagenbeck has been made into something that he was not, and only when we are willing to consider alternative views of Hagenbeck will progress be made in understanding the meaning and importance of his legacy. Günter Niemeyer, for example, writes that in 1957 the idea was circulated of putting together an anniversary people show. Plans moved forward with the cooperation of the Indian Consul in Hamburg to engage a group of "devil dancers, fakirs, snake charmers, and all manner of tricksters." According to Niemeyer, "the Indians were to have been contracted as performers for five months, would have presented their shows daily, and would then have been flown home again without charge."[16] In the end, the Indian embassy in Bonn refused the plan. Seeking an explanation, Niemeyer writes, "Some head or other, it appears, still continued to be haunted by the propaganda story— brought up in both wars—that Hagenbeck 'had, for money, shown natives from foreign lands in the zoo, just as if he were a slave trader.'"[17] It is probably true that propaganda about Hagenbeck's people shows could have adversely affected the company's plan to launch a new show nearly twenty-six years after the last exhibit focusing on the South Seas had closed so inauspiciously. But given the history of the shows and the more recent history of World War II and the Holocaust, important reasons beyond simple propaganda underscore why representatives of India would not look favorably upon a new exhibit of "devil dancers, fakirs, snake charmers, and all manner of tricksters."

"Where do the fish come from?" A simple question such as this about the origins of the fish in the many public and private aquariums in this country helps put Hagenbeck's animal business in perspective. It seems on the whole to be fairly easy today for people to condemn a figure such as Hagenbeck, but if we look around a little, it is not that difficult to see that, in most respects having to do with the exhibition of animals and people, the world has not changed that much since Hagenbeck's time. If the popularity of people shows seems to have diminished, we need not look far to see their continued presence in our culture, not simply on the pages of various geographical magazines and in the videos of nature programs but on the daily talk shows, where all manner of human oddity is presented for mass consumption. To be sure, certain animals in zoological gardens are rarely caught in the wild anymore. There seems, for example, to be plenty of certain species of zebra, deer, antelope, and lions produced in zoos. Indeed, it has even become necessary from time to time for zoos to find

outside buyers for surplus animals. At the same time, international regulations regarding the capture, sale, and transport of many species have pushed zoological gardens to become significantly more creative in their efforts to maintain captive stocks. Among the more publicly popular of these efforts, of course, are the "species survival programs" through which zoos work cooperatively to breed endangered species. The results of these programs reap a double benefit for the zoos: more rare animals for the collections, and more press about how zoos are working toward the conservation of species, biodiversity, and other laudable goals. Despite debate about certain actions of certain gardens, however, many zoos and aquariums are clearly making substantial contributions to preserving and conserving the world's natural resources while also attempting to educate the public about complicated issues surrounding animal keeping and conservation. The prominent role of gardens, for example, in literally preserving from extinction such species as the California condor and the black-footed ferret has shown well that zoos can play important roles in conservation efforts.

Nevertheless, nagging questions continue to arise, such as, "Where do the fish come from?" The National Aquarium in Baltimore, the Monterey Bay Aquarium in northern California, and the New England Aquarium in Boston are part of some very progressive thinking about how aquariums and zoos can educate people, aid science, actively contribute to conservation and preservation programs, and provide needed recreation. But only recently and still only rarely do we hear discussion of a series of other issues including how all those marine fish—of which a mere handful of species have been bred successfully in captivity—end up in the aquarium in the first place. Where *do* they come from? Who catches them? How many die in the process? Are the methods used in catching those fish damaging to the environment? How long do the various species live in the aquarium before new representatives need to be obtained? Do large public collections such as these encourage people to set up their own saltwater aquariums, thus expanding the market for exotic fish?[18] These issues are lost, if not carefully obscured, in the overwhelming onslaught of information about all the wonderful contributions, including even urban renewal, which public aquariums are making to the world, particularly in their efforts to save marine mammals.

I bring forward these criticisms of contemporary popular aquariums—while avoiding the simpler targets of institutions such as Sea World—in order to suggest to those who would too easily dismiss a figure such as Carl Hagenbeck for caring more about money than animals and people that they pause and consider their own interactions with the exotic animal business, including watching *Animal Planet* on cable and *Zoboomafoo* on public television. After years of studying Hagenbeck, there is no doubt in my mind that even if he was occasionally uncomfortable around the members of his people shows, he truly loved being around animals. Do I doubt that the lioness Triest was somehow deeply fond of him and even went into battle for him against her grottomates? Not really. Nor do I doubt that he would have done his best for her or any of his other animals. Nevertheless, as the firm of Carl Hagenbeck passed its 150th anniversary in 1998—celebrated with special postage stamps, special exhibits, and a new official biography—it seemed happy to continue to advance a particularly sanguine vision of Carl Hagenbeck and his firm. To be sure, there is a great deal today for the company and Carl Hagenbeck's descendants to look back on with pride. Carl Hagenbeck built a company from an ancillary activity begun by his father into a business recognized around the world. He was an acquaintance of the leading scientific and political figures of his day both in Germany and around the world. He was as comfortable corresponding with William Temple Hornaday and Baron Walter von Rothschild, as he was entertaining P. T. Barnum and Thomas Alva Edison. When we look at a photograph of the sixty-five-year-old Hagenbeck walking beside Kaiser Wilhelm II during one of the emperor's visits to the park, when we look carefully at Hagenbeck's broad stride and natural manner, and when we consider the beginning of his company half a century earlier, we should not fail to be impressed. With this said, however, it is time for us to consider as well the lives of the animals and people who were brought back alive to Hagenbeck's. This book has tried to tell some of their stories.

Notes

Introduction *Entering the Gates*

1. Alexander Sokolowsky, *Beobachtungen über die Psyche der Menschenaffen* (Frankfurt: Neuer Frankfurter Verlag, 1908), 17–18. For the ease of readers, the first reference in the body of the text (i.e., excluding the notes and bibliographic essay) to a work with a non-English title is translated parenthetically. Subsequent references to that title will be in the original. The titles of popular newspapers and journals, however, have generally not been translated.

2. Ibid., 21.

3. Ibid., 21–22.

4. This assessment—one with which, incidentally, many animal catchers and zookeepers agreed—did not, however, discourage Sokolowsky about the desirability of continuing to capture and import gorillas; rather, he encouraged catchers and importers to catch as yet unweaned young and raise them in captivity in Africa for longer periods before attempting to bring them to Europe (ibid., 22). In her 1915 *From Jungle to Zoo* (New York: Moffat, Yard, 1915), an account of how animals arrive at zoological gardens, Ellen Velvin writes of the deaths of gorillas in captivity: "There seems to be no particular ailment from which they are suffering; no cold or fever, nothing but intense home- or heart-sickness; and there is no doubt whatever to my mind that they grieve themselves to death. . . . They will sit, with their shoulders hunched up, their knees under their chins, and their hands either hanging listlessly down in front of them, or else held to their heads, as though suffering from headache, and never lift their sad and weary eyes for hours at a time. When they are finally induced to look up, there is an expression in their eyes which haunts one, so immeasurably sad and forlorn is the hopelessness of it" (90–91).

5. Sokolowsky, *Beobachtungen*, 51. The comparison Sokolowsky has in mind here is with the chimpanzee.

6. Paul B. Du Chaillu, *Explorations and Adventures in Equatorial Africa; with accounts of the manners and customs of the people, and of the chace of the Gorilla, Crocodile, Leopard, Elephant, Hippopotamus, and other animals* (New York: Harper, 1861), 70.

7. Ibid., 70–71.

8. I use the word *wild* with caution because most of the animals discussed in this book, though often associated with the "wild," spent most of their lives in thoroughly human environments. At the same time, of course, I want to be careful about describing those sup-

posedly "really wild" spaces of the earth which often turn out to be far from our expectation for that term.

9. Theodore Roosevelt, *Hunting Trips of a Ranchman: Sketches of Sport on the Northern Cattle Plains* (New York: G. P. Putnam's Sons, 1886), 261.

10. John Berger, "Why Look at Animals?" in *About Looking* (London: Writers and Readers, 1980), 22.

11. Ibid., 19.

12. Richard Nelson, *Heart and Blood: Living with Deer in America* (New York: Vintage, 1998), xi.

Chapter 1 *Gardens of History*

1. Among the cities to which Clara was brought were Amsterdam, Leiden, Haarlem, Hamburg, Hanover, Berlin, Frankfurt/Oder, Breslau, Vienna, Munich, Regensburg, Freiburg, Leipzig, Cassel, Frankfurt/Main, Mannheim, Bern, Zurich, Basel, Schaffhausen, Strasbourg, Stuttgart, Augsburg, Ansbach, Nuremberg, Würzburg, Leiden, Paris, Lyon, Marseilles, Naples, Rome, Bologna, Verona, Venice, London, Warsaw, Danzig, and Copenhagen. The literature on the rhinoceros is very extensive. The most exhaustive annotated bibliographic source is Leendert Cornelis Rookmaaker's *Bibliography of the Rhinoceros: An Analysis of the Literature on the Recent Rhinoceroses in Culture, History, and Biology* (Rotterdam: Balkema, 1983). T. H. Clarke's *The Rhinoceros from Dürer to Stubbs: 1515–1799* (New York: Harper and Row, 1986) is a remarkable study of the rhinoceros in the visual arts.

2. Dürer's woodcut was based on a letter and sketch by Valentim Fernandes of a rhinoceros that had recently been brought from India to the court in Lisbon.

3. Clarke, 58–60.

4. For a brief discussion, see Immanuel Wallerstein, *The Modern World-System I: Capitalist Agriculture and the Origins of the European World-Economy in the Sixteenth Century* (Orlando, Fla.: Academic Press, 1974), 173–221.

5. Friedrich Knauer, *Der Zoologische Garten: Entwicklungsgang, Anlage und Betrieb unserer Tiergärten* (Leipzig: Theod. Thomas Verlag, n.d. [ca. 1911]), 113.

6. For examples, see Gustav Loisel, *Histoire des Menageries* (Paris: O. Doin et fils, 1912), and C. V. A. Peel, *The Zoological Gardens of Europe: Their History and Chief Features* (London: F. E. Robinson, 1903).

7. J. Vosseler, "Die Wissenschaftliche und Volksbildende Bedeutung eines Zoologischen Gartens im Herzen der Großstadt," address presented at the Hamburg St. Georges Verein of 1874 on May 1, 1911.

8. David Friedrich Weinland, "Ueber den Ursprung und die Bedeutung der neueren Zoologischen Gärten," *Der Zoologische Beobachter* (1862): 43.

9. Dr. Heini Hediger, "Changes in Zoological Gardens," *Anthos: Landscape Architecture Quarterly* 3 (1971): 2–4.

10. For typical pieces of public relations for American zoological gardens which have appeared in contexts that are widely perceived as unbiased, see Cliff Tarpy, "New Zoos—Taking Down the Bars," *National Geographic* (July 1993): 2–37, and *Animal Attractions: Amazing Tales from the San Diego Zoo,* which appeared on the public television *Nature* hosted by George Page (WNET, New York, 1997). For earlier accounts of the new exhibits in gardens around the world, see articles by H. Hediger, C. R. Schmidt, R. E. Honegger, K. Brägger, F. Vogel, and L. Dittrich in the special issue of *Anthos: Landscape Architecture Quarterly* 3 (1971), and articles by G. R. Jones, J. C. Coe, R. Kirchshofer, D. Beisel, M. Bushing, and B. v. Tscharner in the special issue of *Garten und Landschaft: Journal for Landscape Architecture and Landscape Planning* (January 1985).

11. Linda Koebner, *Zoo Book: The Evolution of Wildlife Conservation Centers* (New York: Forge, 1994), 60. Admittedly, these two paragraphs do not really do the book much justice. While its history may lack a critical edge, the book is a beautifully illustrated and very informative introduction to the key individuals and institutions at the forefront of American zoo practice.

12. R. J. Hoage, Anne Roskell, and Jane Mansour, "Menageries and Zoos to 1900," in *New Worlds, New Animals: From Menagerie to Zoological Park in the Nineteenth Century,* ed. R. J. Hoage and William Deiss (Baltimore: Johns Hopkins University Press, 1996), 15.

13. Jacob Burckhardt, *The Civilization of the Renaissance in Italy* (1860; New York: Harper, 1958), 287–88.

14. Ibid., 288 n. 2. Burckhardt notes that in the pitting of the animals against each other in the piazza, "the lions lay down and refused to attack the other animals." In the end, Burckhardt admits that his brief discussion represents "only fragments of a great subject" (292).

15. John Berger, "Why Look at Animals," in *About Looking* (London: Writers and Readers, 1980), 19.

16. Keith Thomas, *Man and the Natural World: A History of the Modern Sensibility* (New York: Pantheon, 1983), 277. Thomas also points to the gardens serving an "aesthetic satisfaction."

17. Ritvo notes, for example: "In thriving menageries the defining symbolism was embodied in every aspect of the exhibits, as well as expressed in public pronouncements and policy formulation. It was implicit in the very composition and structure of a collection. Animals that roamed free and often dangerous in their native wilds were confined in small cages and placed along well-marked paths, in manicured parks that seemed natural only in contrast to the surrounding urban landscapes. . . . Zoos explicitly encouraged visitors to enjoy the contrast between the wild beasts and their intensely cultivated surroundings. See Harriet Ritvo, *The Animal Estate: The English and Other Creatures in the Victorian Age* (Cambridge: Harvard University Press, 1987), 217–18.

18. For information on the beginning of the Jardin, see Loisel, and Jules T. Hamy, "The Royal Menagerie of France and the National Menagerie established on the 14th of Brumaire of the Year II (November 4, 1793)," *Annual Report of the Board of Regents of the Smithsonian Institution, Showing the Operations, Expenditures, and Condition of the Institution to July 1897* (Washington, D.C.: Government Printing Office, 1898), 507–17.

19. David Blackbourn and Geoff Eley, *The Peculiarities of German History: Bourgeois Society and Politics in Nineteenth-Century Germany* (Oxford: Oxford University Press, 1984), 200.

20. Hediger is a notable exception in this camp. In his works, including *Wild Animals in Captivity: An Outline of the Biology of Zoological Gardens*, trans. G. Sircom (London: Butterworths, 1950), and *Man and Animal in the Zoo: Zoo Biology*, trans. Gywnne Vevers and Winwood Reade (New York: Delacorte, 1969), Hediger is consistently battling with what he terms pathologic behaviors in both humans and animals.

21. See, for example, the description in Philip Lutley Sclater, *Guide to the Gardens of the Zoological Society of London* (London: Zoological Society of London, 1870).

22. For a provocative account of the idea of the "pet," see Yi-Fu Tuan, *Dominance and Affection: The Making of Pets* (New Haven: Yale University Press, 1984).

23. See Simon Schama, *Landscape and Memory* (New York: Knopf, 1995), 37–53.

24. Colin Tudge, *Last Animals at the Zoo: How Mass Extinction Can Be Stopped* (Washington, D.C.: Island, 1992), 55.

25. Johann Basilii Küchelbecker, *Allerneueste Nachricht von Romisch-Kayserl. Hofe* (Hanover, 1730).

26. Ibid., 789–91.

27. Ibid., 786.

28. Ibid., 784.

29. Hans Aurenhammer, *The Belvedere in Vienna: Ten Engravings and a Description by Contemporaries of Prince Eugene* (Vienna: Osterreichische Galerie, 1963), 2. For more on the Belvedere menagerie, see Loisel, 67–68.

30. Derek McKay, *Prince Eugene of Savoy* (London: Thames and Hudson, 1977), 193.

31. In the Küchelbecker view of the gardens, these collections can be seen to the right of the main garden. At the upper end of this smaller garden can be seen a wire aviary for indigenous birds.

32. For a good comparison, see Wilfried Hansmann, *Gartenkunst der Renaissance und der Barock* (Cologne: Dumont Buchverlag, 1983), esp. 79–88 and 207–27.

33. McKay, 197.

34. It seems that the Belvedere takes the famous menagerie at Versailles one step further. Designed by Le Vau, the menagerie at Versailles also solidly posits the centrality of the observer. Utilizing five sides of an octagon for animal enclosures (the other three being used for entrances and caretakers' buildings), the menagerie contained seven courts by dividing the symmetrically central side (of the five) into three. For the human audience, Le Vau

designed a two-story axial pavilion, the petit château. It was decorated simply and had a small gallery for paintings of the animals (again suggesting a continuity of the inside and outside), and visitors walking along the balcony could successively observe various groupings of the collection, with only a little effort. This architectural obsession with the viewer, however, is promoted with even greater vigor in Kleiner's view of Eugene's menagerie. Whereas from Le Vau's château the seven animal courts had to be engaged individually by a moving spectator (the views sweep about 225 degrees as one walked around the château), Eugene's courts constitute a much smaller portion of a circle and allow the prince to stand still at the center of his collection.

35. McKay, 197–98.

36. Max Braubach, *Prinz Eugen von Savoyen: Eine Biographie. Band V, Mensch und Schicksal* (Munich: R. Oldenbourg Verlag, 1965), 65.

37. Küchelbecker, 793.

38. Braubach, 63.

39. Henry Scherren, F.Z.S., *The Zoological Society of London: A Sketch of Its Foundation and Development* (London: Cassell, 1905), 7.

40. Ibid., 20.

41. Ibid., 7.

42. *The Zoological Gardens, a Description of the Gardens and Menageries of the Zoological Society: A Handbook for Visitors* (London: H. G. Clarke, [ca. 1860–62]), 5–6. This guide was one of the many unofficial guides to the gardens. The society's "official guides" were published at irregular intervals from 1829 to 1857, when a standard format was adopted under the secretariat of D. W. Mitchell.

43. John P. Haines to William Hornaday, September 4, 1903. Incoming Correspondence, Director's Office, Archives of the New York Zoological Park, Wildlife Conservation Society.

44. The importance of feeding is also evident in the A. A. Milne's "At the Zoo," in which the voice of Christopher Robin proclaims: "There are lions and roaring tigers, and enormous camels and things,/There are biffalo-buffalo-bisons, and a great big bear with wings,/There's a sort of a tiny potamus, and a tiny nosserus too—/But I gave buns to the elephant when I went down to the Zoo!" A. A. Milne, "At the Zoo," in *When We Were Very Young* (London: Methuen, 1924), 46–48. Of course, much of the feeding activity remains crucial to zoos today, although it is now considered completely inappropriate for visitors to the gardens to participate actively in the feeding accept when being instructed in the correct methods by an animal keeper. Still, elements of the older practice of feeding remain important in some zoos. At my last visit to the Berlin Zoo (in the pre-reunified city), for example, the feeding of the lions was a spectacular event in which the lions were brought to a frenzy of excitement before being thrown large chunks of meat publicly hacked off a carcass by a keeper. Of course in the United States, where lions are typically fed a highly processed and enriched meat tray, such a scene would be very much out of place.

45. P. J. S. Olney, "The Policy of Keeping Birds in the Society's Collections, 1826–1976," in *The Zoological Society of London, 1826–1976 and Beyond,* ed. Lord Zuckerman, Symposium of the Zoological Society of London, Number 40 (London: Zoological Society of London, 1976), 134–35.

46. For a fascinating account of bird cage design through the centuries, see Norbert Humburg, *Alte Vogelbauer: Ein Brevier* (Braunschweig: Klinkhardt und Biermann, 1965).

47. The ongoing controversies surrounding the exhibition of pandas suggest that this is as much the case today as it was in the nineteenth century. The most important critic of the exhibition of pandas at zoological gardens in recent years has been George B. Schaller, director of science for the New York Zoological Society / The Conservation Society. See Schaller's *The Last Panda* (Chicago: University of Chicago Press, 1993).

48. See Phillips Verner Bradford and Harvey Blume, *Ota: The Pygmy in the Zoo* (New York: St. Martin's, 1992).

49. Friedrich Katt, "Hagenbecks Tierparadies," *Zoologische Beobachter* 50 (1909): 371.

50. Katt, 372.

51. Ludwig Zukowsky, *Carl Hagenbecks Reich: Ein deutsches Tierparadies* (Berlin: Wegweiser, 1929), 9.

Chapter 2 *Catching Animals*

1. Franz Kafka, "A Report to an Academy," trans. Willa and Edwin Muir, in *Franz Kafka: The Complete Stories,* ed. Nahum N. Glatzer (New York: Schocken, 1971), 250. Although the debate over the evolution of humans from apes does play a role in the "Report," that debate, along with the issue of the lives of performing apes, provides only a general framework for Kafka to discuss his larger concern about the artist and his unappreciative and misinterpreting public. Thus, like Kafka's "The Hunger Artist," even though the "Report" is based on the facts of a certain kind of exhibit—the trained ape that drinks, smokes, shakes hands, spits, and so on—of which there were many examples in the day, this story is not, in its most important sense, about an ape.

2. Ibid., 252.

3. Carl Zuckmayer, "Hamburg, Hafen, Hagenbeck," *Arche Noah* 1 (1949): 3. As quoted in Herman Reichenbach, "Carl Hagenbeck's Tierpark and Modern Zoological Gardens," *Journal of the Society for the Bibliography of Natural History* 9 (1980): 573, and Günter H. W. Niemeyer, *Hagenbeck: Geschichte und Geschichten* (Hamburg: Hans Christians, 1972), 7.

4. The idea that the seal exhibition in 1848 represents the beginning of Hagenbeck's *Tierhandel* has been the official story of and by the company since at least the 1880s. Whether March 1848 is an entirely accurate or useful date to mark the beginning of the company—the seals were shown as mermaids in much the same way that Carl Hagenbeck's

father exhibited all manner of livestock "freaks" during the period—we should recognize that the date has a certain additional appeal for the company.

5. Heinrich Leutemann, who was acquainted with the elder Hagenbeck, named Carl Hagenbeck's father, Claus Carl Gottfried, in his *Lebensbeschreibung des Thierhändlers Carl Hagenbeck* (Hamburg: Carl Hagenbeck, 1887); Niemeyer in his *Hagenbeck: Geschichte und Geschichten* refers to the father as Carl Gottfried Claus. Meanwhile, Erna Mohr in her article "Das Geschlecht Hagenbeck," *Der Zoologische Garten* 21, no. 1 (1954): 2–9, refers to him as Gottfried Clas Carl Hagenbeck. I have followed Mohr's order and Herman Reichenbach's good advice to refer to the elder Hagenbeck, the father of Carl Gottfried Heinrich Hagenbeck, as Carl Hagenbeck Sr. (1810–87) and to the son as Carl Hagenbeck (1844–1913).

6. Carl Hagenbeck, *Von Tieren und Menschen: Erlebnisse und Erfahrungen* (Leipzig: Paul List, 1908), 31.

7. Although Carl had taken over most of the control of the business by 1860, he became the sole proprietor of Carl Hagenbeck's Handelsmenagerie in 1866.

8. Among the other English dealers should be noted Charles Rice, William Cross, and, later, J. D. Hamlyn. Other German traders were competing in this market, as well, including August Fockelmann in Hamburg and the families of Reiche and Ruhe in Alfeld, the latter companies originating in the explosion of the canary trade by midcentury. For a brief treatment of the various dealers in Germany, including Fockelmann, Reiche, Ruhe, Menges, Christiane Hagenbeck, and Heinrich Möller, see Ludwig Heck, *Heiter-ernste Lebensbeichte: Erinnerungen eines alten Tiergärtners* (Berlin: Im Deutschen Verlag, 1938), 243–57; see also Richard Müller, "100 Jahre Tiergroßhandlung L. Ruhe/Alfeld," *Der Zoologische Garten*, n.s., 28, nos. 2–3 (1963): 71–79. As much as Carl Hagenbeck may have been in competition with these other companies, especially the British ones, it is important to recognize that all the companies had occasion to work closely with one another. The London dealer Charles Rice, in fact, was married to one of Carl Hagenbeck's sisters, and letters between Charles Jamrach and the Hagenbecks—for example, those from Dietrich to "Charles" from Suez on March 3, 1872, and from Zanzibar on March 14, 1873, and that from Carl on November 8, 1877—suggest decidedly friendly and supportive relations. Letters among the Miscellaneous Papers, Hagenbeck Archive, Stellingen.

9. C. Hagenbeck, 58–59.

10. Thanks to Herman Reichenbach for alerting me to the Jamrach connection to the property on Spielbudenplatz discovered by Lothar Dittrich and Annelore Rieke-Müller. See Dittrich and Rieke-Müller, *Carl Hagenbeck (1844–1913): Tierhandel und Schaustellungen im Deutschen Kaiserreich* (Frankfurt: Peter Lang, 1998).

11. In his *Lebensbeschreibung,* Heinrich Leutemann notes that the illustration and accompanying article resulted in a steep climb in the number of visitors coming to Hagenbeck's business (19).

12. For more about Adams and his bears, see Joanne Carol Joys, *The Wild Animal Trainer in America* (Boulder, Colo.: Pruett, 1983), 10–11.

13. C. Hagenbeck, 60.

14. Ibid., 64.

15. Ibid.

16. Ibid., 66. According to the bill, handwritten on a small piece of paper and dated June 19, 1870, Casanova's shipment was sold to Hagenbeck for £2,750 and consisted of some 128 animals—excluding small birds and milk goats and including 9 giraffes, 5 elephants, 2 buffalos, 5 lions, 4 leopards, 2 cheetahs, 25 hyenas, 15 monkeys, 12 cranes, and so on. Miscellaneous Papers, Hagenbeck Archive, Stellingen.

17. C. Hagenbeck, 74.

18. Ibid., 77.

19. In a letter to me of November 4, 1998, Herman Reichenbach states that Lothar Schlawe had corrected an earlier statement by Reichenbach in "Carl Hagenbeck's Tierpark," 574, and by Heinz-Georg Klös in *Von der Menagerie zum Tierparadies: 125 Jahre Zoo Berlin* (Berlin: Haude and Spener, 1969), 65, that Hagenbeck's single shipment doubled the species and tripled the specimens.

20. Hans Hermann Schomburgk, an elephant hunter who later became an animal catcher for Hagenbeck and, later still, a wildlife photographer, published his adventures in many books, including *Wild und Wilde im Herzen Afrikas* (Berlin: Fleischel, 1910); *Bwakukama: Fahrten und Forschungen mit Büchse und Film im unbekannten Afrika* (Berlin: Deutsch-Literarisches Institut, 1922); *Der Kleine Jumbo, das Schicksal eines afrikanischen Elefantenbabys* (Hannover: A. Weichert, 1952); *Von Mensch und Tier und etwas von mir* (Berlin: H. Wigankow, 1947); and *Zelte in Afrika. Fahrten-Forschungen-Abenteuer in sechs Jahrzehnten* (Berlin: Verlag der Nation, 1960). Ole Hansen was a Norwegian walrus hunter who had been active in the Arctic since 1886 and who became Hagenbeck's main source for walruses beginning around 1906.

21. For further information on Schillings, see Ludwig Heck, "Die Schillingsche Sendung Deutsch-Ostafrikanische Tiere," *Die Gartenlaube* (1900): 562–64; C. G. Schillings, *Flashlights in the Jungle,* trans. Frederic Whyte (New York: Doubleday, 1905); C. G. Schillings, *Der Zauber des Elelescho* (Leipzig: R. Voigtländer, 1906); and C. G. Schillings, *In Wildest Africa,* trans. Frederic Whyte (New York: Hutchinson, 1907).

22. The works of Hans Dominik include *Kamerun. Sechs Kriegs- und Friedensjahre in deutschen Tropen* (Berlin: Mittler, 1901) and *Vom Atlantik zur Tschadsee. Kriegs- und Forschungsfahrten in Kamerun* (Berlin: Mittler, 1908).

On Carl's half brother John, see especially John Hagenbeck, *John Hagenbecks abenteuerliche Flucht aus Ceylon. Meine Ausweisung aus Ceylon und Flucht nach Europa* (Dresden: Deutsche Buchwerkstätten, 1917); *Fünfundzwanzig Jahre Ceylon. Erlebnisse und Abenteuer im Tropenparadies,* ed. Victor Ottmann (Dresden: Deutsche Buchwerkstätten, 1922); *Kreuz und Quer durch die Indische Welt. Erlebnisse und Abenteuer in Vorder- und Hin-*

terindien, Sumatra, Java, und auf den Andamanen, ed. Victor Ottman (Dresden: Deutsche Buchwerkstätten, 1922); *Südasiatische Fahrten und Abenteuer. Erlebnisse in Britisch- und Holländisch-Indien, im Himalaya und in Siam,* ed. Victor Ottman (Dresden: Deutsche Buchwerkstätten, 1924); the pamphlet *Gebr.-Hagenbeck's Indische Carawane* (Hamburg: n.p., n.d), Jacobsen Papers, Hamburgisches Museum für Völkerkunde; and Wilhelm Munnecke, *Mit Hagenbeck im Dschungel* (Berlin: Scherl, 1931).

23. For more about Casanova, see, in addition to Carl Hagenbeck, *Von Tieren und Menschen,* and the works of Leutemann and Niemeyer, H. Dorner, "Casanova und Hagenbeck," *Die Gartenlaube* 69 (1869): 42–47, and Wilhelm Fischer, *Aus dem Leben und Wirken eines interessanten Mannes* (Hamburg: Baumann, 1896). Beyond the biographies of Hagenbeck and the histories of the company, little can be found about Bernard Cohn (also known as Bernhard Kohn). See, in addition, however, "Afrikanische Thier-Expedition," *Die Gartenlaube* 76 (1876): 844, and *Carl Hagenbeck's Thier-Karawane aus Nubien* (Hamburg: n.p., n.d.), Jacobsen Papers, Hamburgisches Museum für Völkerkunde. A number of letters can be found pertaining to Josef Menges in the Hagenbeck Archive, the Jacobsen Papers of the Hamburgisches Museum für Völkerkunde, and various zoo archives, such as the Archives of the Bronx Zoo. Menges also made regular contributions between 1881 and 1891 to *Petermanns Mittheilungen* about the geography of northeastern Africa. See also *J. Menges' Ost-Afrikanischen Karawane aus den Somalilande* (Hamburg: n.p., n.d.), Jacobsen Papers, Hamburgisches Museum für Völkerkunde.

24. The works of Christoph Schulz include *Auf Großtierfang für Hagenbeck: Selbsterlebtes aus afrikanischer Wildnis* (Dresden: Verlag Deutsche Buchwerkstätten, 1921) and *Jagd- und Filmabenteuer in Afrika: Streifzüge in das Innere des dunklen Erdteils* (Dresden: Verlag Deutsche Buchwerkstätten, 1931). His wife, Elisabeth Schulz, published a volume of "fiction" about their adventures in Africa entitled *Afrikanische Nächte: Erzählung* (Hamburg: Zoo Verlag [Christoph Schulz], 1926), the first (and apparently only) of a projected series to be edited by Christoph Schulz under the series title Im dunkelsten Erdteil: Reiseerzählungen aus Afrika. For further information on Johannsen and Grieger, see C. Hagenbeck and Niemeyer.

25. Schomburgk arrived in South Africa with one thousand marks to start a farm but spent the money immediately, he tells us, to outfit an expedition to the interior.

26. The relatively high capital outlays involved in this business, combined with the conservative purchasing patterns of the directors of the European and American zoological gardens (these were people who liked to deal with "reputable" companies), help explain why Menges, despite years of effort, did not succeed in offering a competitive alternative to Hagenbeck. Although by the 1890s he was sending out printed price lists for animals he had brought back from North and East Africa, he found it difficult to sell his stock more cheaply than Hagenbeck and consistently complained in correspondence that, although he could bring many more animals back, he found it difficult to find buyers for them. In a letter to Johan Jacobsen of July 27, 1894, for example, Menges writes that although he had just

returned from three months in Somalia, having brought back a great many animals, "unfortunately the sale of the animals has gone poorly, even worse than last year. The gardens have no money and do not want to buy anything. The last lot of animals are dying miserable deaths instead of being sold, the Berlin and Dresden gardens, for example, haven't spent a penny." Referring to the year before, in which he had lost a substantial amount of money at the World's Fair in Chicago, where he was trying to sell ethnographic artifacts, Menges laments, "If this goes on, this year will be even worse than the last in Chicago, in which I was completely devastated" (Josef Menges to Johan Adrian Jacobsen, July 27, 1894, Jacobsen Papers, Hamburgisches Museum für Völkerkunde). Given the strength and "market share" of the firm of Carl Hagenbeck, it is simply hard to imagine that Menges could compete in the large animal business in the 1890s in Germany. Only a company with a clearly established presence in the national and international market, such as that of Ludwig Ruhe in Alfeld, could hope to take significant business away from Hagenbeck in the large animal market.

27. In his introduction to C. G. Schillings' popular version of his *Mit Blitzlicht und Büchse im Zauber des Eleléscho* (Leipzig: Voigtländer, 1910), Theodore Roosevelt described Schillings as "a mighty hunter and explorer, a trained field naturalist, an exceedingly interesting writer, and the most noted of nature-photographers." "His book," Roosevelt concluded, "should be circulated everywhere and translated into every language spoken by those men who love the wilderness, and the beasts of the wilderness, and the wild, hardy life of the biggame hunter."

28. For a description of the Schillings hunt, see Schillings, *Mit Blitzlicht,* 110. For descriptions of the Hagenbeck expeditions to Mongolia, see, among others, C. Hagenbeck, 212–28.

29. Based on internal evidence, the pamphlet *Carl Hagenbeck's Thier-Karawane aus Nubien* was most likely published in 1876 and referred to a shipment organized by Bernard Cohn. This pamphlet, along with others, including *J. Menges' Ost-Afrikanische Karawane aus dem Somalilande* (Hamburg, n.p., n.d.), and some private correspondence were used to complete sections in Hagenbeck's *Von Tieren und Menschen.* The whole story in Hagenbeck's memoir of how the Hamran hunters captured elephants with swords, for example, is taken from *Carl Hagenbeck's Thier-Karawane aus Nubien.* Copies of these brochures can be found in the Jacobsen Papers at the Hamburgishes Museum für Völkerkunde.

30. *Carl Hagenbeck's Thier-Karawane,* 11.

31. Ibid., 12.

32. Ibid., 14. The "Nubian Caravan" organized by Bernhard Cohn and brought back by Hagenbeck in 1876 was exhibited in Hamburg, Düsseldorf, Breslau, and Paris. The exhibit attracted more than thirty-thousand visitors in a single day in Breslau. Another "Nubian Caravan," that of 1877–78, again organized by Cohn, traveled to Hamburg, Berlin, Frankfurt, Dresden, and London. Josef Menges also organized a "Nubian" show in 1878. For more information about the Hagenbeck people exhibitions, see Chapter 3 below and Hilke

Thode-Arora, *Für fünfzig Pfennig um die Welt: Die Hagenbeckschen Völkerschauen* (Frankfurt: Campus, 1989).

33. Josef Menges, "Bemerkungen über den deutschen Thierhandel von Nord-Ost-Afrika," *Der Zoologische Garten* 17 (1876): 231–32.

34. It should be noted that Menges had an interest in making the trade seem easier than it probably was. With Hagenbeck controlling more and more aspects of the business and Menges apparently hoping to become independent of the Hagenbeck operation, it was important for Menges to present the procuring of animals as a relatively easy affair so that he could convince the generally risk-averse zoological gardens to underwrite his future plans.

35. See Endre Stiansen, "Overture to Imperialism: European Trade and Economic Change in the Sudan in the Nineteenth Century" (Ph.D. diss., University of Bergen, 1994).

36. For a more complete discussion of the credit systems in West Africa, see Colin Newbury, "Credit in Early Nineteenth Century West African Trade," *Journal of African History* 13, no. 1 (1972): 81–95.

37. Menges lists the prices of animals in Kassala in Maria Theresia talers (1 taler = 4 marks). In addition to the prices for elephants, giraffes, and rhinoceroses, Menges notes that an antelope could be purchased in Kassala for 20–120 marks and sold in Europe for 400–3,000 marks and a young lion could be purchased for 8–80 marks and sold for 600–2,400 marks. See Menges, "Bemerkungen," 232.

38. Figures given by Menges, in ibid., 233.

39. I have not simply imagined this shipment. Hagenbeck's 1870 Casanova/Migoletti importation included, in fact, one rhinoceros, five elephants, fourteen giraffes, twelve antelopes, and seven young lions. It further included two warthogs; four aardvarks; four wild Nubian buffalo; fifty-three large and small beasts of prey, excluding the seven lions; twenty-six African ostriches, among which were sixteen full-grown birds; and twenty large crates with monkeys and birds. The "Nubian Caravan" described above represented an even more valuable shipment, consisting of three rhinoceroses, fourteen elephants, nine giraffes, fifteen antelopes, and seventeen ostriches.

40. See the stories reported by Ellen Velvin, *From Jungle to Zoo* (New York: Moffat, Yard, 1915), 185, 80.

41. Leutemann, 33–34.

42. It is quite possible that Casanova and the other early traders occasionally caught animals themselves. All the information about the early hunters suggests, though, that they rarely ventured after animals and relied on indigenous hunters, who, naturally, would have had an idea about which animals were of most interest to the company—at the time, elephants, giraffes, and ostriches. Dietrich Hagenbeck's attempt to catch hippos remains the earliest instance of the company sending out a hunter after a particular animal. To give a sense of the relative value of hippopotamuses in the period, an 1883 price list from Hagenbeck shows a six-year-old female hippo as the most expensive animal in Hagenbeck's col-

lection, selling for 18,000 marks, followed by a seven-year-old male African rhinoceros for 12,000 marks.

43. See Diederich Hagenbeck to Charles [Jamrach], March 14, 1873, Hagenbeck Archive, Stellingen.

44. Dietrich Hagenbeck to Carl Hagenbeck, April 29, 1873, published as correspondence under the title "Ein junger Nilpferdjäger: Dietrich Hagenbeck," *Die Gartenlaube* (1873): 754.

45. Dietrich Hagenbeck to Carl Hagenbeck, June 25, 1873, *Die Gartenlaube* (1873): 754.

46. Thanks to Edward Steinhart for identifying "black water fever."

47. Dietrich Hagenbeck to Carl Hagenbeck, June 25, 1873, 754.

48. For discussion of Dominik's earlier animal collections, see *Vom Atlantik*. Dominik relates the Heck dialogue in *Kamerun*, 240.

49. Both Hans Schomburgk and Carl Gustav Schillings noted Dominik's accomplishment with envy. Schillings, also at the instigation of Heck, tried repeatedly to catch and then keep alive a baby elephant from German East Africa. In each case in which he succeeded in catching a calf, it soon died, however. See Schillings, *Mit Blitzlicht*, 55–56.

50. Dominik, *Kamerun*, 242.

51. Ibid., 244.

52. Ibid., 248.

53. Ibid., 252.

54. Some question remains about how many of the young elephants survived over the longer term. Dominik writes that three survived, but Schillings notes that only one animal survived—the one that eventually arrived in Berlin. By the end of this episode in his life, Dominik had become convinced that Europeans could look forward to a future in which tamed elephants could be used in Africa as they were in India. He concluded: "In discussions of the thousands of tons of ivory which are shipped yearly from Africa and more specifically from the Congo, one hears time and again the anxious question, when will the last elephant in Africa be shot? Even if it will be a good while until then, so it would be greeted with great satisfaction, if the last wild elephant were to be carried to his grave by his domesticated brothers" (*Kamerun*, 253).

55. Leutemann, 29.

56. Schomburgk, *Wild*, xiii. An exception that he notes was the necessity of killing a mother elephant in his eventually successful attempt to procure a baby elephant for the zoological gardens in Berlin. During his career as an elephant hunter, Schomburgk killed sixty-three elephants (see Schomburgk, *Von Mensch*, 119–20).

57. Ibid., 121.

58. Ibid., 277.

59. Ibid., 278.

60. Ibid. In response to a question about the outline around Jumbo, Jutta Niemann, the

granddaughter of Schomburgk and a photo editor, confirmed that the picture seems to have been enhanced so that the contrast between Jumbo and his mother is clearer.

61. See Christoph Schulz, *Auf Großtierfang*. Once the connection to the Hagenbecks was made, Schomburgk mounted special expeditions for the company, the most important of which was his successful first capture and transport to Europe of the pygmy hippopotamus from Liberia in 1912. The account of this adventure is described in detail in Hans Hermann Schomburgk, *Pulsschlag der Wildnis* (Berlin: Verlag der Nation, 1959). See also "Hagenbecks Zwergflußpferd," *Zoologischer Beobachter* 53 (1912): 132–34, 208–10, 338–39, and discussion below.

62. Schomburgk, *Wild*, 319. The claim that Jumbo was the first East African elephant to reach Europe is inaccurate. In any case, Jumbo was the first elephant from the colony of German East Africa to arrive in Germany. Schomburgk's interest in Jumbo continued long after their reunion in Stellingen. Some forty years after the appearance of his *Wild*, Schomburgk published a rather poor but, in this context, interesting children's story centering on the adventures of the young elephant. Partly lifted from his earlier memoir, partly imagined, and partly combined with material unrelated to Jumbo or his capture, *Der kleine Jumbo* is more a curiosity than a work of historical or literary quality. This is not the place to go into depth about this book, but it is important to note that, as with most of the stories in this chapter, the "fate" of "little Jumbo" as it was painted in Schomburgk's stories was a mixture of pain over the past and joy and excitement about a new adventure in a different land. But even in a children's story, the central fact of the killing of Jumbo's mother was, however unpleasant, simply inescapable. The best that the hunter could do in his memoirs and memory, it seems, was to describe the confrontation with and death of the mother—despite his intentions from the beginning—as being somehow completely unavoidable, as almost an accident, and to show that Jumbo's new life was certainly as good as, if not better than, his old one.

63. Schomburgk, *Von Mensch*, 39.

64. Schomburgk writes, for example, "Nature doesn't recognize sentimentality. If an animal is driven out, it is meant to be. In most cases, a sick animal will leave its herd of its own will. Animals understand how to die without making a fuss. The inner natural instinct teaches them that their individual lives are meaningless, but that their death can be useful to the community, that their death serves the species. This is so because sick or weak animals can bring the downfall of the entire herd. As in a chain, which is also only as strong as its weakest link, the weakest member of an animal community must always be at least strong enough to meet fully all the demands of nature" (*Von Mensch*, 110).

65. On the death of Major Philip Pretorius, who had killed some 555 elephants during his career, Schomburgk referred to "English statistics" (figures he habitually brought up in his books) which suggested that an elephant hunter could expect to live only two years before being trampled by an elephant, murdered by the natives, fatally bitten by a snake, or perishing from malaria, black water, or some other tropical disease. Returning to Pretorius,

Schomburgk writes: "555 elephants! Who can possibly understand what that means! Hunters will marvel at this Nimrod, 'only' friends of animals will damn him. I say 'only' intentionally, because every true hunter is also a friend of animals, and by the 'only' I refer to those enthusiastic and sensitive friends of animals who are not hunters and who, in their admirable eagerness, carelessly overshoot their target" (*Von Mensch,* 121).

66. Elizabeth Hanson in her 1996 dissertation, "Nature Civilized: A Cultural History of American Zoos, 1870–1940" (University of Pennsylvania, Philadelphia), on the American animal trade and American zoological gardens, insightfully describes Frank Buck, one of the most famous animal catchers of the twentieth century, not as a heroic animal catcher per se—especially since he himself rarely participated in the actual capture of the animals—but rather as a type of heroic businessman who struggles, under the most adverse conditions, to get his animals to market.

67. William T. Hornaday to Carl Hagenbeck, June 11, 1902, Outgoing Correspondence, Director's Office, New York Zoological Park, Wildlife Conservation Society.

68. See Renate Hücking and Ekkehard Launer, *Aus Menschen Neger machen: Wie sich das Handelshaus Woermann an Afrika entwickelt hat* (Hamburg: Galgenberg, 1986), 87. In this regard, Dominik seems to be a good match to Cameroon's governor at the time, Jesko von Puttkamer—a man sympathetic to the needs of German merchants for increased control in the colony who described the Duala in his memoirs as "the most lazy, false, and base rabble the sun ever shone upon" and stated further: "It would certainly have been better if, during the conquest of the land in 1884, they had at least been kicked out of the country if not exterminated" (87). See also Albert Wirz, *Vom Sklavenhandel zum kolonialen Handel: Wirtschaftsräume und Wirtschaftsformen in Kamerun vor 1914* (Zurich: Atlantis, 1972), 37, and Jesko von Puttkamer, *Gouverneursjahre in Kamerun* (Berlin: G. Stilke, 1912), 51.

69. Helmuth Stoecker, in his *Kamerun unter Deutscher Kolonialherrschaft,* Band I (Berlin: Rütten and Loening, 1906), provides a fairly gruesome catalogue of atrocities visited upon the indigenous peoples of Cameroon. See, especially, 196–221.

70. Quoted in Hücking and Launer, 89–90.

71. Hücking and Launer excerpt, for example, a 1907 article in the newspaper *Kolonie und Heimat:* "The honest citizen is overcome by a slight horror when he reads about corporal or chaining punishments in the colonies. He is inclined to regard them as relics of a sinister barbarism and the perpetrators who are forced to use these forms of punishment as inhuman" (88). The article, however, goes on to explain that although these types of punishment are naturally abhorred by civilized people both at home and in the colonies, they remain the only truly efficacious methods to discipline indigenous peoples.

72. Josef Menges, for example, describes the Somalis as "unbathed, shameless, brazen, and presumptuous savages" in his letter to Johan Jacobsen of March 30, 1891. Jacobsen Papers, Hamburgishes Museum für Völkerkunde.

73. See "On the Trail of the Pygmy Hippo: An Account of the Hagenbeck Expedition

to Liberia," *Zoological Society Bulletin* 16, no. 52 (1912), and "At Last New York Has a Pair of Pygmy Hippos," *New York Times*, July 14, 1912, Magazine Section, pt. 5, p. 14.

74. Signed unpaginated typescript sent by Hans Schomburgk to William Hornaday and housed with the Director's Incoming Correspondence at the New York Zoological Society Archives. The differences between the manuscript and the version printed in the *Zoological Society Bulletin* are minor; the version printed in the *New York Times* is substantially edited.

75. My thanks to Harold James for bringing to my attention the history of the Steiff stuffed animal business. The success of the various manufacturers of stuffed animals in the late nineteenth century, of which the German company of Steiff is only the most prominent, is, of course, deeply connected to the growth of zoological gardens in the period. That, as John Berger pointed out, "the manufacture of realistic animal toys coincides, more or less, with the establishment of public zoos" can be extended, however, to include a number of other concurrent sentimental developments including the literary/scientific controversy known as the "nature faker" debate that raged in North America at the turn of the century. Focusing on highly anthropomorphized natural historical writing, the debate featured John Burroughs and Theodore Roosevelt on one side and William Long and Ernest Thompson Seton on the other in an often furious discussion about the oversentimentalization of animals in works of natural history. See John Berger, "Why Look at Animals?" in *About Looking* (London: Writers and Readers, 1980), and Ralph H. Lutts, *The Nature Fakers: Wildlife, Science, and Sentiment* (Golden, Colo.: Fulcrum, 1990).

76. Colonel Richard Meinertzhagen, *Kenya Diary, 1902–1906* (Edinburgh: Oliver and Boyd, 1957), 178. What makes this condemnation so notable is that Meinertzhagen was arguably one of the most "blood-thirsty" British officials in the history of Kenya.

77. Friedrich Knauer, *Der Zoologische Garten: Entwicklungsgang, Anlage und Betrieb unserer Tiergärten* (Leipzig: Theod. Thomas Verlag, n.d. [ca. 1911]), 117–18.

78. See Christoph Schulz, *Aus Hagenbecks Jagdgründen: Abenteuer eines Tierfängers in den Steppen und Urwäldern Afrikas* (Leipzig: Verlag Deutsche Buchwerkstätten, 1922). See also Schulz's separate volume entitled *Jagd- und Filmabenteuer,* which focuses on one adventure during the 1910–13 period described in *Auf Großtierfang* when he shot some nine thousand meters of film in Africa to be shown in theaters, including that at the Stellingen Animal Park.

79. Schulz, *Auf Großtierfang,* 9.

80. Despite few literary aspirations, Schulz's work does pause for moments of more poetic description. They are usually quickly lost, however, as the author returns to the more predictable. In an account of a particularly striking morning, for example, we read: "I was astonished at the beginning of dawn by a spectacular natural display. All was still covered in darkness, then I saw in the far distance the two snow-covered summits of Kilimanjaro, the Kibo and Mawenzi, light up in the first rays of the climbing sun. As if spellbound I stood and enjoyed the splendid scene. Gradually it became lighter and the outline of the

huge and imposing mountain became more and more clearly visible. Over a wide area the two summits, which jutted high above a layer of clouds, glimmered. For the first time I saw the giant of the African mountains and my heart rejoiced in bliss over the overpowering and magical splendor which this view offered to me. Over there, then, lay German territory, and with pleasure I greeted the king of the German East-African mountain world, which the preceding day had been covered in mist" (*Auf Großtierfang*, 78–79).

81. Carl Hagenbeck to William T. Hornaday, February 3, 1913, Incoming Correspondence, Director's Office, New York Zoological Park, Wildlife Conservation Society. Hagenbeck typically corresponded with his English-speaking clients in English—even if his language was sometimes awkward.

82. Schulz, *Auf Großtierfang*, 90.

83. Ibid., 92.

84. Ludwig Zukowsky, *Carl Hagenbecks Reich: Ein deutsches Tierparadies* (Berlin: Wegweiser, 1929), 62.

85. Ibid., 234. It is important, however, to note that although Zukowsky argues that the Hagenbeck family rules insist that, in the effort to catch animals, killing other animals should at all costs be avoided, Heinrich and Lorenz Hagenbeck had stepped up the importation of the walruses, receiving ten in 1928; and Zukowsky himself notes that the method of catching the animals was virtually unchanged since the nineteenth century. He writes: "After a mother with a catchable young is sighted, the old one is either driven onto the land or harpooned from the ship and then killed with a shot to its head; the young one is then caught with a lasso. Touching scenes take place between the dying mother walrus and her baby. Until the last moment the old lady attempts to bring her baby to safety. Because the young walrus itself does not leave its dead mother, it is easy to pull it by its flippers on board or catch it with a lasso from the boat . Young walruses become accustomed relatively easily to fish for food, so it is not especially difficult to keep the valuable animals alive" (148–49).

86. C. Hagenbeck, 414.

87. John Hagenbeck, like his brother, was always ready, it seems, to put his not inconsiderable talents into diverse enterprises. According to his own account, he first went to Ceylon in 1884 to organize an exhibition of people (*Völkerausstellung*) for his brother, although some confusion remains about which, if any, show he helped to organize (see, for example, Thode-Arora, 46). After various activities for the company, including assisting brother-in-law Heinrich Mehrmann with the Hagenbeck circus and helping with various other *Völkerausstellungen*, John settled in Colombo. His first business interests centered around the pearl fishery, but there he met with little success (*Kreuz und Quer*, 10). Frustrated in the pearl business, John Hagenbeck took a position with a ship chandlery and before long was turning his hand to a wide variety of activities. Of primary interest throughout remained the catching and shipping of exotic animals to Europe and America, but he remained consistently involved in organizing *Völkerausstellungen*, as well. With his brother Gustav, moreover, he founded an exporting company concerned chiefly with tea. He fled

Ceylon with the outbreak of World War I, leaving his wealth and possessions behind, and worked beside his nephew Lorenz in Stellingen for a short while before becoming involved in a movie house in Berlin. He first returned to Ceylon in 1920, traveling there more regularly beginning in 1924 to organize more people exhibitions. According to Thode-Arora, he returned with his family to Ceylon in 1927, regaining a position in the ship chandlery and reestablishing himself as a planter. He died in British internment during World War II on December 16, 1940.

88. See note 22 above.

89. Munnecke notes, for example, that John Hagenbeck went to India in 1885 to catch animals, it seems, and that after World War I he did not return to the Indian subcontinent before May 1929.

90. Munnecke, 9.

91. Ibid., 15.

92. Ibid., 58–59.

93. Ibid., 17.

94. Ibid., 18.

95. Ibid., 19–20.

96. Ibid., 23.

97. Munnecke tells a longer story of the most famous of the three young tigers, Radja. Despite being a completely trustworthy performer, and indeed a sensation since he was never shown behind bars, Radja one day accidentally killed a monkey. Having tasted blood and having thereby heard the atavistic "call of the wild," Radja, the Indian prince, could no longer be trusted by the circus owners and was forced to live out his days behind bars. The motif here of the Indian prince who can only partly be civilized is taken up elsewhere in the work.

98. Munnecke, 93.

99. Ibid., 78–79.

100. Kafka, 251.

Chapter 3 *"Fabulous Animals"*

1. Wilhelm Fischer, *Aus dem Leben und Wirken eines interessanten Mannes* (Hamburg: Baumann, 1896), 24.

2. Carl Hagenbeck, *Von Tieren und Menschen: Erlebnisse und Erfahrungen* (Leipzig: Paul List, 1908), 93.

3. A number of the words used by Germans and Europeans more generally to connote various indigenous peoples in the nineteenth century have been rightly displaced. The words *Lap* and *Eskimo,* for example, were commonly used by Hagenbeck and his representatives. I have avoided the use of these words, except in quotations or in the titles of the

exhibits. I have, however, retained more neutral names of countries or regions in describing these people (e.g., Sudanese, Greenlander, or Laplander) without quotation marks.

4. C. Hagenbeck, 94.

5. Ibid., 96.

6. Ibid.

7. Hagenbeck generally used two terms to describe these shows: *Völkerschau* ("people show") and *Völkerausstellung* ("exhibition of people").

8. C. Hagenbeck, 96.

9. Ibid., 99.

10. Ibid.

11. Ibid., 100.

12. Ibid., 101.

13. Ibid., 100. As is not entirely uncommon, Hagenbeck was not quite accurate in this statement. See William C. Sturtevant and David B. Quinn, "This New Prey: Eskimos in Europe in 1567, 1576, and 1577," in *Indians in Europe: An Interdisciplinary Collection of Essays,* ed. Christian Feest (Aachen: Herodot, 1987), 61–140.

14. C. Hagenbeck, 103.

15. Ibid., 121.

16. Ibid.

17. Ibid., 127.

18. Ibid., 96 (my emphasis).

19. L. Beckmann, "Die Hagenbeck'schen Singhalesen," *Die Gartenlaube* (1884): 564.

20. Alexander Sokolowsky, *Carl Hagenbeck und sein Werk* (Leipzig: E. Haberland, 1928), 64.

21. Christopher Columbus, *Das Bordbuch 1492. Leben und Fahrten des Entdeckers der neuen Welt in Dokumenten und Aufzeichnugnen,* ed. Robert Grün (Tübingen: Erdmann, 1974), 139. Quoted in Stefan Goldmann, "Wilde in Europa: Aspekte und Orte Ihrer Zurschaustellung," in *Wir und die Wilden: Einblicke in eine Kannibalische Beziehung,* ed. Thomas Theye (Reinbek bei Hamburg: Rowohlt, 1984), 243.

22. Urs Bitterli, *Die "Wilden" und die "Zivilisierten": Grundzüge einer Geistes- und Kulturgeschichte der europäisch-überseeischen Begegnung* (Munich: Beck, 1976), 180.

23. Goldmann, 265 n. 2; Bitterli, 180–81.

24. Sturtevant and Quinn, 69.

25. Ibid.

26. From his first voyage, Frobisher returned with an Inuit man (originally taken in the hope of exchanging him for lost members of the ship's crew), who died some fifteen days after arriving in England. On his second voyage, Frobisher returned with three captives— a man, Kalicho, and a woman and her child, Arnaq and Nutaaq—whose legacy in oral and visual sources in England is remarkably extensive. See Sturtevant and Quinn, 68–88.

27. Bitterli, 181–87.

28. C. Hagenbeck, 95.

29. Heinrich Leutemann, *Lebensbeschreibung des Thierhändlers Carl Hagenbeck* (Hamburg: Carl Hagenbeck, 1887), 48.

30. According to the reports in the *Verhandlungen* (Transactions) of the Berliner Gesellschaft für Anthropologie, Ethnologie und Urgeschichte (Berlin Anthropological Society), the group consisted of eight individuals, the twenty-three-year-old man Klemme and his mother and father, Aennta and Hennta; two other young men, Dovit and Jona, both twenty-six years old; and three women, Karim, who was eighteen years old, and Ippa and Kaisa, who were thirty-two and thirty-four years old, respectively. See Rudolf Virchow, "Zur Vorstellung der nach Berlin gebrachten Lappen," *Verhandlungen der Berliner Gesellschaft für Anthropologie, Ethnologie und Urgeschichte* 7 (1875): 225–28.

31. When the article was written, only four members of the eight-member troupe had arrived in Berlin. The woman referred to is Karim.

32. *Die Neue Preußische (Kreuz-) Zeitung* (Berlin), January 20, 1875.

33. Heinrich Leutemann, "Lappländer: Nordische Gäste," *Die Gartenlaube* 75 (1875): 742.

34. C. Hagenbeck, 96.

35. I am indebted to Suzanne Marchand for her discussion of verisimilitude in her paper "Of Words and Things: The Antimonies of Vorgeschichtsforschung," Panel #50, German Studies Association Annual Conference, 1993. Marchand argues pointedly for the importance that verisimilitude gained in studies of prehistory—a point I turn to at more depth below—because of the central problems of sources and professional legitimacy faced by scholars in the field.

36. *Frankfurter Zeitung und Handelsblatt,* 2nd morning ed., June 7, 1889. According to the June 18, 1889, evening edition of the paper, during its time in Frankfurt, the Somali exhibit, which was organized by Hagenbeck's associate Josef Menges, was attended by some 30,500 adults and 10,400 schoolchildren.

37. I am using the Hagenbeck shows here as exemplars of a type because, on the one hand, other exhibits organized by different impresarios clearly had the same goals and, on the other, several Hagenbeck shows fell short of his ideal of verisimilitude. Among the shows in the first category, we should count the Somali shows organized by Josef Menges; among the shows in the latter, we should include the 1878 show of East Indians composed of a group of servants from England, which Hagenbeck found very disappointing, and the 1910 Oglala-Sioux show, composed of forty-two Indians and ten cowboys, which clearly mimicked the type of popular western show made famous by Buffalo Bill. For a fairly complete list of people shows organized by both Hagenbeck and his competitors in Germany, see Hilke Thode-Arora, *Für fünfzig Pfennig um die Welt: Die Hagenbeckschen Völkerschauen* (Frankfurt: Campus, 1989).

38. Sokolowsky remembered, "It was none other than Rudolf Virchow in Berlin, the great anatomist and anthropologist, who conducted thorough measurements and exami-

nations of the human material brought to him [by Hagenbeck], and then made his results available to science through publication." Sokolowsky, *Carl Hagenbeck*, 66.

39. One of the recurring frauds exposed by the scientific establishment, for example, was the exhibit of the "Aztec Children." Shown throughout Europe from the early 1850s through to the first decade of the twentieth century, this exhibit became a repeated focus for Rudolf Virchow. In response to the claim that these were the last survivors of a vanished race, Virchow sought to prove that the alleged "Aztecs" were, in fact, only suffering from microcephalism. See Nigel Rothfels, "Aztecs, Aborigines, and Ape-People: Science and Freaks in Germany, 1850–1900," in *Freakery: Cultural Spectacles of the Extraordinary Body*, ed. Rosemarie Garland Thomson (New York: New York University Press, 1996), 158–72.

40. Since his death in 1902, many studies have attempted to assess the influence that Rudolf Virchow exerted over the scientific, political, medical, and larger popular thought of several generations. For the major works and complete bibliographies, see Erwin H. Ackerknecht, *Rudolf Virchow: Doctor, Statesman, Anthropologist* (Madison: University of Wisconsin Press, 1953); Christian Andree, *Rudolf Virchow als Prähistoriker* (Cologne: Bohlau, 1976–86); and Manfred Vasold, *Rudolf Virchow: Der Große Arzt und Politiker* (Stuttgart: Deutsche Verlags-Anstalt, 1988). Ackerknecht's biography divided Virchow's accomplishments into three main fields: his efforts as a medical doctor, as a political figure, and as an anthropologist. As a physician, Virchow is perhaps most famous for his articulations of the cell theory, *omnis cellula a cellula* (each cell stems from another cell), and the theory of cellular pathology. Working from the 1837 cell theory of Theodor Schwann, who, while recognizing that the cell was the basic unit of life nevertheless believed that it coagulated from an amorphous "blastema," Virchow, who was studying the development of disease in the body, advanced in 1854 his eventual theory that all cells derive from other cells; a year later, his groundbreaking essay "Cellular Pathology" contained the aphorism *omnis cellula a cellula*. To be sure, the theory had been suggested earlier by other scientists, but none had ever given up completely on Schwann's ideas of "free cell formation" (Ackerknecht, 70–84). Despite Robert Remak's actual precedence in elaborating this theory, Virchow reached his conclusions independently, and it is from Virchow's discussion of the cell and his articulation of the theory of cellular pathology—that cells of disease derive from healthy cells—that the medical community finally accepted the cellular theory. Ackerknecht writes, "Working always almost exclusively on full-fledged pathological material, he reached his conclusions later [than Remak]; but they had far more meaning and immediate applicability for the medical profession. In addition to the fact that Virchow's findings were more meaningful, he propagated them with that tireless zeal and almost sinister energy in which nobody ever excelled him" (84).

As a political figure in nineteenth-century Germany, Virchow is only somewhat less estimable. Beginning with his efforts to combat a typhus epidemic in Upper Silesia in 1847–48, Virchow showed himself to be a political liberal and activist for social reform. Convinced that the epidemic stemmed as much as anything else from the inadequacy of the outmoded feudal political structure of the region, Virchow—whose prescription for Silesia included full and unlimited democracy, freedom, the promotion of education, self-government, the use of the Polish language, the separation of church and state, and increased taxes for the wealthy—returned to Berlin in mid-March 1848, just as the revolution hit. His response to the Revolution of 1848 was elation. Instrumental in the formation of the Medical Reform Movement in 1849, Virchow maintained a high political profile and helped forge the German Progressive Party (Deutsche Fortschrittspartei) in 1861 out of an assortment of democrats, liberals, and moderates, becoming a key spokesman for the party during the constitutional conflict of 1862–66.

As an anthropologist, Virchow is perhaps best known for his leading roles in the debates over the authenticity of the Neanderthal Man and Heinrich Schliemann's discovery of Troy, but his less well known systematic studies of the physical anthropology of various races—including his massive survey of the racial characteristics and variability of German schoolchildren in 1876—clearly occupied his thoughts more than the popular debates about Troy and cavemen. In the end, it seems that on almost any matter, from such public health issues as sewage to debates over Darwinian evolution (which he actively opposed because it carried with it an element of the old arguments about spontaneous generation), Virchow's ideas were likely to be solicited as often by the educated elites of Boston, Paris, and London as by their German counterparts.

41. Virchow, "Über Eskimos," *Verhandlungen der Berliner Gesellschaft für Anthropologie, Ethnologie und Urgeschichte* 10 (1878): 185. Not surprisingly, Virchow's lecture was reported in the papers. See, for example, *Die Neue Preußische (Kreuz-) Zeitung,* March 13, 1878.

42. The evening edition of *Frankfurter Zeitung und Handelsblatt* of July 21, 1896, reported, "Yesterday the specialist in indigenous shoes [Naturschuhwerk], Herr Eberhard Müller, completed his measurements of the feet of the beautiful and strong Samoans." The results of the examination, which the paper recorded in brief, evidently led Herr Müller to the conclusion that the "relatively broader separation" of the toes of the Samoans in comparison with the Europeans was the result of the Samoans' walking barefoot more frequently.

43. The declining importance of Hagenbeck's exhibits of people to the Berlin Anthropological Society (not Berlin or German culture more generally, however) can be not completely coincidentally dated to the end of Virchow's prominent role in the society in the late 1890s. In the first decade of the twentieth century, only two references are made to people shows in the society's *Verhandlungen,* namely, a Chinese group of 1905 and a "Tscherkessen-Troupe" organized by Eduard Gehring and seen at the Berlin Zoological Gardens in the spring of 1900. Both references are brief and suggest more simple curiosity than an earnest scientific endeavor. Virchow's own interest in the shows only partly explains their fall from the society's favor, however; of perhaps more importance was the evolving significance of ethnographic studies being completed in the field, including such

early examples as those published by Gustav Fritsch as *Die Eingeborenen Südafrika's: Ethnographisch und Anatomisch Beschrieben* (Breslau: F. Hirt, 1872) and by Bernhard Hagen as *Unter den Papua's: Beobachtungen und Studien über Land und Leute, Thier- und Pflanzenwelt in Kaiser-Wilhelmsland* (Wiesbaden: C. W. Kreidel, 1899).

44. Interestingly, in press reports, the Christian names of the members of this group were generally excluded from the copy. It seems that most observers—and this is also true for Virchow, who throughout the remainder of his report used the "native" names for the people—considered the Christian names inauthentic.

45. Virchow, "Über Eskimos," 187.

46. Ibid.

47. Ibid., 188.

48. See R. Neuhauss, "Die Neuordnug der Photographischen Sammlung der Anthro-pologischen Gesellschaft," *Verhandlungen der Berliner Gesellschaft für Anthropologie, Ethnologie und Urgeschichte* 40 (1908): 95–100.

49. Sokolowsky, *Carl Hagenbeck*, 65.

50. Quoted in *Die Neue Preußische (Kreuz-) Zeitung*, November 21, 1881.

51. *Die Neue Preußische (Kreuz-) Zeitung*, November 21, 1881.

52. Virchow, "Über die Eskimos von Labrador," *Verhandlungen der Berliner Gesell-schaft für Anthropologie, Ethnologie und Urgeschichte* 12 (1880): 270.

53. Ibid.

54. See Virchow, "Über die sogenannten Azteken und die Chua," *Verhandlungen der Berliner Gesellschaft für Anthropologie, Ethnologie und Urgeschichte* 23 (1891): 370–77, and "Über die beiden Azteken," *Verhandlungen der Berliner Gesellschaft für Anthropologie, Ethnologie und Urgeschichte* 33 (1901): 348–50. See also Rothfels, and Hans Scheugl and Felix Adanos, *Show Freaks and Monsters* (Cologne: DuMont, 1974).

55. See Virchow, "Über die in Berlin anwesenden Nubier," *Verhandlungen der Berliner Gesellschaft für Anthropologie, Ethnologie und Urgeschichte* 10 (1878): 333–55, and "Über die ethnologischen Verhältnisse der Nubier," *Verhandlungen der Berliner Gesellschaft für Anthropologie, Ethnologie und Urgeschichte* 10 (1878): 387–407.

56. The tradition on which these measurements was based reaches back at least to eight-eenth-century attempts to divide the races of humans by physical features. In his popular anthropological handbook, *Naturgeschichte des Menschen: Grundriss der somatischen An-thropologie* (Stuttgart: F. Enke, 1904), Carl Heinrich Stratz pointed to the major historical advances in anthropometric analysis. While Carl Linnaeus (1707–87) divided Homo sapi-ens principally by skin color and geography, adding to two categories that were soon dropped (Homo ferus and Homo monstrousus) the divisions Homo europaeus, Homo asiaticus, Homo afer, and Homo americanus, Peter Camper (1722–89) was the first to use the skull as a primary focus for division, developing what became known as Camper's face angle. According to Stratz, one calculated the face angle by drawing a line from the fore-most point on the maxilla to the forehead and another from the same point on the maxilla

to the foremost point of the lower portion of the skull. Thus, among animals, according to Stratz, the angle is very small, and it develops to almost a right angle with humans (Stratz, 4). In 1790 the German Friedrich Blumenbach (1752–1840) divided the races of humans by skulls, coming up with five types, the Mongolian, American, Caucasian, Malaysian, and Ethiopian. The Swede Retzius the Older followed the ideas of Blumenbach and Camper and developed the idea of skull indices, which he applied to all peoples in 1842, coining terms that would be used throughout the century, including *dolichocephalic* (long-headed) and *brachycephalic* (short-headed) and adding two more descriptive terms classifying the jaw structure, *orthognatic* (straight-toothed) and *prognathic* (slant-toothed). In 1861 Paul Broca added the concept of *mesocephalic* (medium-headed) and three terms describing the nose, *leptorrhinic* (narrow-nosed), *mesorrhinic* (medium-nosed), and *platyrrhinic* (broad-nosed), and in 1863 Friedrich Müller, building on the work of Pruner-Bey, categorized different hair types with (1) *ulotriches* (woolly-haired), including *lophokomoi* (tuft-haired) and *eriokomoi* (fleece-haired), and (2) *lissotriches* (simple-haired [Schlichthaarige]), including *euthykomoi* (straight-haired [Straffhaarige]) and *euplokomoi* (locked-haired). Finally, Kollmann in 1881 added *leptoprosopic* (long-faced) and *chamäoprosopic* (short-faced) to the descriptive terms used by anthropologists.

57. According to *Die Neue Preußische (Kreuz-) Zeitung* of November 3, 1880: "Today Professor Virchow undertook his anthropological measurements among the Eskimos currently residing at the zoological garden. The following was reported: Mrs. Págna [Bairngo], who is known among her people as a sorceress, observed Virchow's investigations with visible distrust from the very beginning. As she herself became the object of the study, she made clear that she thought Virchow was a sorcerer of the gods of the white men and that he had come to steal her powers. In the middle of the examination, she suddenly sprang up and jumped about like an ape over the tables and chairs, uttering spells in gurgling tones all the while. Professor Virchow and the two men assisting him retreated, and only after a good deal of time was it possible to calm Mrs. Pánga again."

58. Virchow, "Über die Eskimos von Labrador," 271.

59. Of all the measurements, the identification of skin color and tone was the least exact. The tables often did not have a good match (see, for example, the investigation of the "Nubians" in 1878), and anthropologists were uncertain where the measurement should be taken because the tone varied across the body.

60. Virchow, "Über die ethnologischen Verhältnisse der Nubier," 403.

61. Stratz, 256.

62. Ibid.

63. Hermann Heinrich Ploss and Max Bartels, *Woman: An Historical, Gynaecological, and Anthropological Compendium* (1913; London: Heinemann, 1935), 395–456.

64. See discussion in Jan Lederbogen, "Fotografie als Völkerschau," *Fotogeschichte. Beiträge zur Geschichte und Aesthetik der Fotografie* 6, no. 22 (1986): 47–64.

65. I have taken the idea of "achievement" from Stratz's introduction to the concept of

Ethnographie (1) because it demonstrates, again, the essentially comparative framework of these studies. To limit confusion, I have used the slightly less common term *ethnology* to encompass the German terms *Ethnologie* and *Ethnographie*. In general, the two terms were used interchangeably by members of the Berliner Gesellschaft für Anthropologie, Ethnologie und Urgeschichte, that is, both terms pointed to the various efforts to describe and compare races and cultures without the use of the sciences of archaeology and physical anthropology. While our current use of the term *ethnography* focuses more on the description rather than the comparison of cultures, and while there were instances of late-nineteenth-century German *Ethnographie* which seem quite close to our understanding of the concept, the discipline had much more in common with what we term *ethnology*, which encompasses all aspects of the description and comparisons of cultures. Thus researchers of indigenous religions and languages in the Berlin Anthropological Society were usually drawn into the larger questions of the origins and distributions of races and cultures, and their studies were, as a rule, incomplete without the authors' having addressed the "more significant" questions raised by the cultures they studied.

66. I use the quotation marks around the word *donate* because the items were usually sold to the museums to help cover the expenses of the shows. Hagenbeck's carefully maintained connections to the Berlin Anthropological Society, to which he was voted a member in the fall of 1878 (during the first Nubian show), and the Hamburg Ethnographic Museum, of which he was one of the founding contributors, often facilitated these purchases.

67. Leutemann, *Lebensbeschreibung,* 50.

68. Josef Menges to Johan Adrian Jacobsen, January 11, 1891, Jacobsen Papers, Hamburgishes Museum für Völkerkunde.

69. The best book to date on Jacobsen is Hilke Thode-Arora's 1989 *Für fünfzig Pfennig um die Welt.* Thode-Arora based her book, the first scholarly monograph dedicated to the Hagenbeck people shows, on the Jacobsen Papers in the Hamburgisches Museum für Völkerkunde. In addition to the strength of her original research, Thode-Arora deserves particular praise for her unwillingness to jump to simple conclusions about the importance of the shows.

70. Thode-Arora, 50.

71. For accounts of some of these adventures, see, for example, Otto Genest, "Kapitän Jakobsen's Reisen im Lande der Golden," *Globus. Illustrirte Zeitschrift für Länder- und Völkerkunde* 52, no. 10 (1887): 152–56; 11:171–74; 13:205–8; 14:220–23; or Jacobsen's own "A. Jacobsen's und H. Kühn's Reise in Niederländisch-Indien," *Globus. Illustrirte Zeitschrift für Länder- und Völkerkunde* 55, no. 11(1889): 161–68; 12:182–86; 13:200–204; 14:213–17; 15:225–29; 16:244–48; 17:261–65; 18:279–80; 19:299–302; *Captain Jacobsen's Reise an der Nordwest-Küste Amerikas, zum Zwecke Ethnolog. Sammlungen,* ed. A. Woldt (Leipzig: M. Spohr, 1884); and *Reise in die Inselwelt des Banda-Meeres,* ed. Paul Roland and intro. Rudolf Virchow (Berlin: Mitscher and Röstell, 1896). Not surprisingly, it appears that members of the Anthropological Society looked down on Jacobsen. Thode-

Arora quotes the editor of Jacobsen's record of his travels on the northwest coast of America, A. Woldt, who noted that Jacobsen "did not travel as a specialist, but as a simple collector and trader. . . . [He] simply purchased and traded for everything that was to be had" (Thode-Arora, 52). Of course, Jacobsen's methods were more practical and opportunistic than "scientific" or systematic. For an account of one of his trading sessions, see Jacobsen, "A. Jacobsen's und H. Kühn's Reise in Niederländisch-Indien," *Globus. Illustrirte Zeitschrift für Länder- und Völkerkunde* 55, no. 14(1889): 183.

72. *Die Neue Preußische (Kreuz-) Zeitung,* August 9, 1884.

73. Ibid. Among the other occasions when the members of the troupes visited the museums, we should note the two visits of Aridji Punchi Banda, the leader of the "Sinhalesen Caravan" of 1884. See the descriptions in ibid., August 17 and 21, 1884.

74. Alexander Sokolowsky, *Menschenkunde. Eine Naturgeschichte sämtlicher Völkerrassen der Erde. Ein Handbuch für Jedermann* (Stuttgart: Union Deutsche Verlagsgesellschaft, 1901).

75. Wolfgang Haberland, "Nine Bella Coolas in Germany," in Feest, *Indians and Europe,* 353–54.

76. Boas nevertheless prepared at least three articles based on his observations of the Hagenbeck show: "Mittheilungen über die Lilxula-Indianer," *Original-Mittheilungen aus der Ethnologischen Abtheilung der Königlichen Museen* (Berlin) 1 (1886): 177–82; "Sprache der Bella-Coola-Indianer," *Verhandlungen der Berliner Gesellschaft für Anthropologie, Ethnologie und Urgeschichte* 18 (1886): 202–6; and "The Language of the Bilhoola in British Columbia," *Science,* 1st ser., 7 (1886): 218.

77. David Friedrich Weinland, *Rulaman. Naturgeschichtliche Erzählung aus der Zeit des Höhlenmenschen und des Höhlenbären* (1878; Stuttgart: Deutsche Verlagsanstalt, 1986), tells the story of the coming of age of a young cave dweller in prehistoric Germany. For a brief description of the life of Weinland, see Hansjörg Küster's afterword to *Rulaman,* 321–35.

78. David Friedrich Weinland, "Gedanken über den Ursprung und das Leben des ureuropäischen Höhlenmenschen," *Die Natur. Zeitung zur Verbreitung naturwissenschaftlicher Kenntnis und Naturschauung für alle Stände,* n.s., 4 (1878): 18.

79. For a provocative account, see the historian Simon Schama's *Landscape and Memory* (New York: Alfred A. Knopf, 1995), esp. 75–134.

80. Stratz, 15. Virchow, it should be noted, was the most important German scientific objector to Darwin's theory of natural selection, insisting that although it seemed probable, it had not and probably never would be proven. See discussion in Vasold's *Rudolf Virchow,* 303–24.

81. Although Darwin recognized in *The Origin* that parallels existed between the ontogeny and phylogeny of an organism, he resisted theorizing a general principle of relation. Of course, by the time of *The Origin* the parallels had been already discussed for decades. In the 1820s, J. F. Meckel, followed by Étienne Serres, elaborated the connections, albeit

within a static (i.e., nonevolutionary) conception of organic life, and in the 1840s and 1850s Louis Agassiz discussed a "three-fold parallelism" with the introduction of the fossil record, noting that the ontogeny of an embryo also repeats fossilized successions. Nevertheless, it is with Haeckel that the biogenetic law "that ontogeny is a concise and compressed recapitulation of phylogeny, conditioned by the laws of heredity and adaptation" gained its overwhelming popularity through, at least, the beginning of World War I. For a clear presentation of the development of the "recapitulation debate," see Ernst Mayr, *The Growth of Biological Thought: Diversity, Evolution, and Inheritance* (Cambridge: Harvard University Press, 1982), 469–76.

82. See discussion in Stephen Jay Gould, *Ontogeny and Phylogeny* (Cambridge: Harvard University Press, 1977), 69–166, and Frank Sulloway, *Freud, Biologist of the Mind: Beyond the Psychoanalytic Legend* (New York: Basic Books, 1979).

83. Weinland, "Gedanken," 3.

84. Ibid., 6.

85. See Marchand, "Of Words and Things."

86. Weinland, "Gedanken," 20.

87. Sigmund Freud, *Totem and Taboo: Some Points of Agreement between the Mental Lives of Savages and Neurotics,* trans. James Strachey (New York: Norton, 1950), 1.

88. The one exception here is the Vedda from Sri Lanka, who, it was frequently asserted, were perhaps the direct ancestors of Aryan peoples. Despite his repeated efforts, however, Hagenbeck never succeeded in bringing a group of Vedda to Germany for exhibition. The importance of the Vedda in the German imagination is suggested by Virchow's monograph *Ueber die Weddas von Ceylon und Ihre Beziehungen zu den Nachbarstämmen* (Berlin: Verlag der Königlichen Akademie der Wissenschaften, 1881).

89. Leutemann, *Lebensbeschreibung,* 62.

90. Ibid.

91. There was a small collection of arrows and stone axes that had been collected in the region. See discussion in Stratz, *Naturgeschichte,* 735. *Die Neue Preußische (Kreuz-) Zeitung* from November 11, 1881, also notes the arrival of a boat from Terra del Fuego, in which, it reports, some of the people were originally found. Most shows came with interpreters or translators. In this case, however, scientists (and the public) were consistently frustrated in their attempts to communicate with the members of the troupe because they made no progress in understanding the language of the Fuegians.

92. *Die Neue Preußische (Kreuz-) Zeitung,* November 18, 1881.

93. Virchow, "Über die Feuerländer," *Verhandlungen der Berliner Gesellschaft für Anthropologie, Ethnologie und Urgeschichte* 12 (1881): 375–93.

94. Ibid., 375.

95. Ibid., 379.

96. Ibid., 381.

97. The names were given by Hagenbeck's agent. No other names for the people were ever learned.

98. Virchow, "Über die Feuerländer," 390.

99. Heinrich Steinitz, in his article "Die Feuerländer" in the popular *Die Gartenlaube* 81 (1881): 732–35, however, noted French articles appearing about the group which agreed with earlier reports that the Fuegians were "repulsive creatures standing still at the very beginning of human culture, if one can even use such a term with them" (732).

100. Virchow, "Über die Feuerländer," 385.

101. This argument, which attempts to understand the status of races within the confines of the physical environment in which they lived, including the nature of the more general flora and fauna of the region, was used especially in cases such as the Inuit, certain Siberian peoples, and the Fuegians, who lived under extreme environmental conditions, but was also applied to explain the "lowness" of the Australian aborigines, which—paralleling, it was argued, the marsupials—had stayed in an arrested stage of development. The argument, which survived well into this century, is typically deployed in Friedrich Raztel's popular *Völkerkunde*. In a review of the work in the bourgeois magazine *Globus* in 1889, we read, for example: "The cultural gap between different groups of humanity is based, according to Ratzel, less on their differing abilities as on the many inner and outer conditions under which a people develops, especially under the distinctiveness of the environment. In the case of the peoples of nature in which practically all effort is spent in the fight for survival, there is then almost no accumulation and production of spiritual achievements" (G. Leipoldt, *Globus* 55, no. 14 [1889]: 210).

102. Observers consistently pointed to the fact that while the Fuegians used a cape around their shoulders to protect themselves from the cold, they apparently made little effort to cover their genitals. Darwin's observations of "Eve's Costume" during his trip aboard the *Beagle* were widely distributed during the show (Steinitz, 734), and Georg Schlesinger reported to the Berlin Anthropological Society that in the low temperatures of Terra del Fuego, "the almost total nakedness of all people, men as well as women, [was] especially noticeable; they had only thrown a skin over their shoulders. A feeling of shame, even among the females, was never apparent" (*Verhandlungen* 12 [1881]: 394).

103. Steinitz argued that cooked food made the people sick; using Virchow's essays extensively, on the other hand, Johannes Ranke in his *Der Mensch* (Leipzig: Bibliographisches Institut, 1887) noted that food was usually cooked: "They roast everything if they can, particularly fish; thereby they also use the fire which they normally take with them in their boats. They do not need any cooking implements for frying; the flesh is laid directly on the ash-covered coals, is turned with a fork shaped rod and when done is then eaten without completely removing the adhering ashes. In any case, they do not think twice about eating meat raw and to gnaw it off the bones" (322).

104. Steinitz, 734.

105. Ranke, 321–22.

106. Heinrich Leutemann, "Die Eskimos in Paris," *Die Natur. Zeitung zur Verbreitung naturwissenschaftlicher Kenntnis und Naturschauung für alle Stände,* n.s., 4 (1878): 115–19, and Weinland's "Gedanken," 3–6; 18–21.

107. Leutemann's specifically literary efforts in his "Die Eskimos in Paris" do not add much to our knowledge of the group. He typically notes the relative smallness of the people, their black straight hair, and comparatively flattened noses and slanted eyes and remarks, "Even according to our own ideas, the 22 year-old [Maggak], despite her slanted eyes, can be acknowledged as beautiful, to which her lively dark eyes, her carefully coifed hair, mild form, and also the red of her cheeks against a fair complexion contribute greatly" (115). Most of the rest of the article, however, attests to the wisdom behind the Danish administration in Greenland, concluding that the government had bestowed on the "Eskimos" the advantages of civilization while limiting its vices. "One cannot deny the Danish government the acknowledgment," wrote Leutemann, "that this is owing to the fortunate combination of intelligence and humanity [in their government] and that they have thereby offered a wonderful example to other European nations" (119). One of the obvious reasons for this discussion was Leutemann's interest (as an agent for Hagenbeck) in keeping good relations with the administrators of Greenland. Despite his efforts, however, the Danish government forbade all future attempts to bring Greenland Inuit to Europe for purposes of exhibition. According to Hagenbeck, the reason behind the Danish decision was that the "Eskimos" returned to Greenland so wealthy that they no longer wanted to work. Just as likely, however, was the government's concern to protect its "charges" from unscrupulous promoters.

108. The work stems from the more than one hundred illustrations that the artist executed for Weinland's *Rulaman* saga and was accompanied in the article by a similar illustration by Leutemann depicting a confrontation between a man and a saber-toothed tiger. Neither of the works in the article, however, was included in the novel, apparently because they did not fit the story.

109. "Gedanken," 20.

110. Ibid.

111. Günter H. W. Niemeyer, *Hagenbeck: Geschichte und Geschichten* (Hamburg: Hans Christians, 1972), 217.

112. Leutemann, *Lebensbeschreibung,* 67.

113. This colonial interest, of course, was not limited to the Hagenbeck shows in Germany or even Europe. A guide for the Chicago Columbian Exposition, for example, noted that "the recent troubles on the island of Samoa, in which the government of the United States intervened, made these people objects of special interest for the visitors at the World's Exposition and their presentations always drew a full house." See *Die Illustrirte Welt-Ausstellung, Chicago, 1893: Das Columbische Weltausstellungs-Album* (Chicago: Rand, McNally, 1893), n.pag.

114. John Hagenbeck, *Fünfundzwanzig Jahre Ceylon. Erlebnisse und Abenteuer im Tropenparadies,* ed. Victor Ottmann (Dresden: Deutsche Buchwerkstätten, 1922), 9.

115. J. Hagenbeck, 10.

116. According to John Hagenbeck in his *Fünfundzwandzig Jahre* (11), the "dressing-up" of Prince Dido became a fascination for the German audiences, and he notes that during the show's run in Berlin, Prince Dido was given no fewer than twelve top hats as gifts. Renate Hücking and Ekkehard Launer, in their study of the firm of Adolf Woermann and its activities in West Africa, note a general assessment among Germans in the late nineteenth century that Africans looked ridiculous in European clothes. See Hücking and Launer, *Aus Menschen Neger machen: Wie sich das Handelshaus Woermann an Afrika entwickelt hat* (Hamburg: Galgenberg, 1986),102–3.

117. J. Hagenbeck, 11.

118. Leutemann, *Lebensbeschreibung,* 68.

119. "Impresario" was a term that the organizers of the shows most often used to describe themselves. See, for example, J. Hagenbeck, 8.

120. Miscellaneous Papers, Hagenbeck Archive.

121. See discussion in Haberland, 337–74.

122. *Gebr.-Hagenbeck's Indische Carawane,* Die Malabaren-Programme (Hamburg: Adolf Friedlander, n.d.). Jacobsen Papers, Hamburgisches Museum für Völkerkunde.

123. For more on this and other posters created for the shows, see Alfred Lehmann, "Zeitgenössische Bilder der ersten Völkerschauen," *Von fremden Völkern und Kulturen: Beiträge zur Völkerkunde. Hans Plischke: Zum 65 Geburtstag gewidmet von seinen Kollegen und Freunden, Schülern und Mitarbeitern,* ed. Werner Lang, Walter Nippold, and Günther Spannaus (Düsseldorf: Droste, 1955); Viktoria Schmidt-Linsenhoff, "Völkerschauen," *Plakate 1880–1914. Inventarkatalog der Plakatsammlung des Historischen Museums Frankfurt,* ed. Viktoria Schmidt-Linsenhoff, Kurt Wettengl, and Almut Junker (Frankfurt: Historisches Museums Frankfurt, 1986); and Ruth Malhotra, *Manege Frei. Artisten- und Circusplakate von Adolph Friedländer* (Dortmund: Harenberg, 1979).

124. The text of program is presented in Hans Werner Debrunner's *Presence and Prestige: Africans in Europe: A History of Africans in Europe before 1918* (Basel: Basler Afrika Bibliographien, 1979), 363–64.

125. Ibid., 364.

126. Peter Altenberg, *Ashantee* (Berlin: Fischer, 1897). Despite various attempts to come to terms with Altenberg and his works, the author remains something of an enigma for modern readers. Although he has assumed almost legendary proportions in Vienna, few readers of German outside Austria will be familiar with his name. In his day, however, Altenberg was widely regarded, especially by other writers, as one of the truly extraordinary authors of his time. Among those contemporaries who outspokenly admired Altenberg should be counted the diverse group of Thomas Mann, Robert Musil, Alfred Kerr, Karl Kraus, Hugo von Hofmannsthal, and Erich Mühsam, while Franz Kafka argued that

Peter Altenberg was "a genius of the trivial, a rare idealist who found the beauties of the world as cigarette butts in the ash trays of coffee houses." See Hans Christian Kosler, *Peter Altenberg. Leben und Werk in Texten und Bildern* (Munich: Matthes and Seitz, 1981), 52.

127. Quoted in Kosler, 82.

128. Altenberg, 29.

129. *Frankfurter Zeitung und Handelblatt,* September 15, 1897.

130. *Die Neue Preußische (Kreuz-) Zeitung,* August 2, 1884.

131. One of Hagenbeck's brothers-in-law, Heinrich Umlauff, built up a large collection of natural objects and wax figures in his *Weltmuseum.* The collection of stuffed animals included a spectacular male gorilla shot by H. Paschen in Cameroon in 1900, which was one of the great gorilla specimens of the time. According to Carl Thinius, "No museum, either within or without the country, owned a specimen of similar proportions." See Carl Thinius, *Damals in St. Pauli: Lust und Freude in der Vorstadt* (Hamburg: Christians, 1975), 35. In addition to the gorilla, any number of animals that had died during their stay at Hagenbeck's Animal Park ended up in the collection. The "Amazon Corps," consisting of fourteen women and ten men, was essentially a fantasy production of the legend of an army of women in Africa; in this case the group was from Dahomey. Again, it seems that Hagenbeck took advantage of a popular story that was widely circulated and, in at least one instance, publicly condemned in a racist attack on "lazy" Africans; see "Die Amazonen von Dahome," *Die Gartenlaube* 88 (1888): 427.

132. Thinius, 36.

133. Ibid., 37.

134. Lederbogen, 55; see also Goldmann, 263, and Thomas Theye, "Eine Reise in Vergessene Schränke. Anmerkungen zu Fotosammlungen des 19. Jahrhunderts in deutschen Völkerkundemuseen," *Fotogeschichte* 5, no. 17 (1985): 14.

135. *Frankfurter Zeitung und Handelsblatt,* 1st morning ed., June 27, 1896.

136. Altenberg, 52–53.

137. Theye, 14.

138. Lederbogen, 55.

139. See Hermann Pollig and Tilman Osterwold, eds., *Exotische Welten, Europäische Phantasien* (Stuttgart: Edition Cantz, 1987). On the significance of the size of the audience for the photographs, Lederbogen notes, "Photos of non-Europeans were available to a larger number of people than the actual shows. Thus, the photographs certainly played a more important role in constructing the European vision of foreigners than their actual physical presence at the exhibits" (61).

140. Ludwig Zukowsky, *Carl Hagenbecks Reich: Ein deutsches Tierparadies* (Berlin: Wegweiser, 1929), 297.

141. Sokolowsky, *Carl Hagenbeck,* 64.

142. Goldmann, 263.

143. It should be reiterated that Virchow was among the few who, for the most part,

avoided speculating on ideas about one race being obviously superior to another. His work, however, when interpreted by other scientists, was easily translated into racist arguments.

144. C. Hagenbeck, 127–28.

145. In the more recent official history of the Hagenbecks from the 1970s, Günter Niemeyer concluded, for example: "Returned to their homes, all the black, brown, yellow, and red people carried with them to their distant Mongolian yurts, Cape corrals (Kaffernkrale), and igloos what for them as well was an extraordinary experience. There, children and grandchildren heard the legends of the great adventure which to this day is bound to the name of Hagenbeck" (217).

146. Niemeyer, 69–70.

147. W. G. Gosling, *Labrador: Its Discovery, Exploration, and Development* (London: A. Rivers, 1910), 309–11. Gosling's patronizing orientation to the Inuit was based, at least, in his deep sympathy for the people. For a remarkably different perspective on the shows of the Labrador Inuit, see J. Garth Taylor, "An Eskimo Abroad, 1880: His Diary and Death," *Canadian Geographic* (October/November 1981): 38–43, which is the story of a diary kept by the troupe member Abraham.

148. When one considers that in one day in Berlin more than ninety-three thousand people paid to see the Kalmucks, then one can begin to have a sense of the potential money involved.

149. An exception seems to have been the efforts undertaken with the extended family of Hersy Egeh, the "chief" of numerous Somali shows. Although his attendance was short lived, for example, one of his sons attended the village school in Stellingen and evidently spoke fluent Platt, and the family, it has been claimed, acquired substantial wealth in Djibouti running a sewing shop using machines brought from Germany. See Niemeyer, 219.

150. Leutemann, *Lebensbeschreibung,* 56–57.

151. C. Hagenbeck, 127–28.

152. Niemeyer, 217.

Chapter 4 *Paradise*

1. Franz Kafka, "A Report to an Academy," trans. Willa and Edwin Muir, in *Franz Kafka: The Complete Stories,* ed. Nahum N. Glatzer (New York: Schocken, 1971), 253.

2. Ibid., 257.

3. Ibid., 257–58.

4. *Daily News* (London), January 23, 1869, quoted in J. W. Toovey, "150 Years of Building at London Zoo," in *The Zoological Society of London, 1826–1976 and Beyond,* ed. Lord Zuckerman, Symposium of the Zoological Society of London, Number 40 (London: Zoological Society of London, 1976).

5. Rainer Maria Rilke, "The Panther, *Jardin des Plantes, Paris,*" in *Neue Gedichte,* trans. Stephen Cohn (1907; Manchester: Carcanet, 1992).

6. Kafka, 253.

7. Carl Hagenbeck, *Von Tieren und Menschen: Erlebnisse und Erfahrungen* (Leipzig: Paul List, 1908), 134.

8. Lorenz Hagenbeck, *Den Tieren gehört mein Herz* (Hamburg: Hoffmann und Campe, 1955), 32.

9. It is because of the Hagenbeck-Wallace Circus more than his exhibitions at the world's fairs in Chicago (1893) and St. Louis (1904) that Hagenbeck's name is remembered in the United States. The Hagenbeck-Wallace Circus, which toured especially the U.S. Midwest into the 1920s, was part of Hagenbeck's operations in name only. From 1902, Carl Hagenbeck's Trained Animal Show Company, and in 1905 the Carl Hagenbeck Circus and Show Company, were managed by an American partnership, and Hagenbeck received none of the profits. The American partners, Frank Tate and John Havlin, eventually sold the circus in 1907 to Ben Wallace, a circus man of mixed repute living in Peru, Indiana. Hearing of the agreement, Hagenbeck cabled Tate and Havlin in January 1907 stating clearly, "I protest and never agree to this." Despite his numerous legal and other efforts over the years, Hagenbeck did not succeed in having his name removed from the operation. For a detailed account, see Fred D. Pfening Jr., "How Wallace Bought Hagenbeck," *Bandwagon* (July/August 1964): 11–12.

10. C. Hagenbeck, 136.

11. Ibid., 139. There is some disagreement about whether Wilhelm or Carl should be credited with the new training techniques and the acts themselves. Carl took credit for them, but his account was published after a severe split in the family. The descendants of Wilhelm tend to credit him with the accomplishment. The circus historians Hans-Jürgen Tiede and Rosemarie Tiede favor Wilhelm on this issue. See their two-part piece "Wilhelm and Willy Hagenbeck," *King Pole* 120 (September 1998): 6–8, and *King Pole* 121 (December 1998): 24–26. My thanks to Richard Reynolds III of the Circus Historical Society for furnishing me with these articles. My own tendency is to give more credit to Wilhelm but to recognize as well that in the early years of the experiments with animal training the two were essentially working together.

12. C. Hagenbeck, 139.

13. Ibid., 143.

14. Ibid.

15. The persecution of the Christians seems to have come from an act prepared by Wilhelm Hagenbeck which was debuted in Paris in 1890. In the act, according to Hans-Jürgen and Rosemarie Tiede, mannequins were fixed to crosses, with meat tied to them, and then set upon by lions. See Tiede and Tiede, 7.

16. *The World's Fair, being a Pictorial History of the Columbian Exposition* (Chicago: n.p., 1893).

17. *Hagenbeck's Arena and World's Museum,* official souvenir program, World's Columbian Exposition, Chicago Historical Society.

18. *Midway Types*. Photo Album. No date, publisher, or pagination. Chicago Historical Society.

19. *Die Illustrirte Welt-Ausstellung, Chicago, 1893. Das Columbische Weltausstellungs-Album*. Chicago: Rand, McNally, 1893. Unpaginated.

20. For more detail on the history of animal training, if largely only in the American context, see Joanne Carol Joys, *The Wild Animal Trainer in America* (Boulder, Colo.: Pruett, 1993). James Gilbert, in his *Perfect Cities: Chicago's Utopias of 1893* (Chicago: University of Chicago Press, 1991), estimates that the German Village on the midway had close to 2 million visitors; the Ferris wheel, at least 1.5 million riders; and Hagenbeck's Pavilion, at least 1 million ticket buyers (122).

21. *Haney's Art of Training Animals: A Practical Guide for Amateur or Professional Trainers* (New York: Jesse Haney, 1869). This is the same work as the *Art of Training Animals: A Practical Guide for Amateur or Professional Trainers. Giving Full Instructions for Breaking, Taming and Teaching all Kinds of Animals. Including an Improved Method of Horse Breaking . . . Serpent Charming; Care and Tuition of Talking, Singing and Performing Birds; and Detailed Instructions for Teaching All Circus Tricks, and Many Other Wonderful Feats* (New York: H. Sample, 1869).

22. This quotation is taken from the May 21, 1903, issue of the *Realm: A Magazine of Marvels*, 60. The dating of the issues changed from week to week, but the content remained stable throughout the season's run.

23. "Notes by a Lion Tamer," *Forest and Stream* December 20, 1883, 408. My thanks to Elizabeth Frank for sending me this article.

24. Cleaveland Moffett, "Wild Beasts and Their Keepers: How the Animals in a Menagerie Are Tamed, Trained, and Cared For." *McClure's Magazine* (May 1894): 552, 554. While we should be skeptical about accepting at face value either the words of Philadelphia or their quotation in *McClure's*, the point is clear that neither Moffett nor Philadelphia appears to have been concerned about depicting training or "taming" as an occasionally violent activity.

25. Ibid., 556.

26. Joys notes that a silver crowbar "magnetized" by the concentration of three German barons was once presented to Van Amburgh by the Drury Lane staff. See Joys, 7.

27. Isaac Van Amburgh, *An Illustrated History and Full and Accurate Description of the Wild Beasts and other Interesting Specimens of Nature, contained in the Grand Caravan of Van Amburgh & Co* (New York: Jonas Booth, 1846).

28. See Joys, 7.

29. Among such figures we should note "Herr Jacob Driesbach, Lord of the Brute Creation" and "Manchester Jack" as well as the "lion queens" Mrs. George Sanger and Helen Bright. The description of Herr Driesbach in the *Hartford Daily Courant* of June 8, 1843, brings forward the idea that the exhibitions could also take place outside a cage. In the article we read, "A green cloth was spread before the cages in the open tent (parlous work, I

thought, among such tender meat as 200 children) and out sprang a full-grown tiger who seized the gentleman [Driesbach] by the throat. A struggle ensues in which they roll over and over on the ground and finally the victim gets the upper hand and drags out his devourer by the nape of the neck. I was inclined to think once or twice that the tiger was doing more than was set down for him in the play, but as the Newfoundland dog of the establishment looked on very quietly, I reserved my criticism."

30. For details on Adams' life, see Joys, 9–11.

31. Raymond Blathwayt, "Wild Beasts: How They Are Transported and Trained," *McClure's Magazine* (July 1893): 132.

32. *Midway Types,* n.p.

33. In an article by Raymond Blathwayt for *Pearson's Magazine* in May 1899 entitled "The Training of Wild Beasts" Hagenbeck is quoted as saying: "My . . . scheme is to inclose large wood and forest spaces in England, or in Europe generally, stock them with wild animals (lions, tigers, panthers, bears, etc.), provide them with food (living animals), take care that they have caves to sleep in and a stream to drink from, and so in course of time sportsmen will have the finest big game shooting in the world at their very gates. It will be perfectly safe for the populace generally, as, of course, the wood will be most carefully inclosed—hermetically sealed I might term it—and for the animals themselves what a boon that they should once again obtain their freedom!" (240–46).

34. Günter H. W. Niemeyer, *Hagenbeck: Geschichte und Geschichten* (Hamburg: Hans Christians, 1972), 254, writes that construction began on twenty-seven hectares. The figure of fourteen comes from Hagenbeck's memoirs; Niemeyer's figure includes various expansions of the park.

35. C. Hagenbeck, 176.

36. Ibid., 173.

37. Ludwig Zukowsky, *Carl Hagenbecks Reich: Ein deutsches Tierparadies* (Berlin: Wegweiser, 1929), 58.

38. C. Hagenbeck, 176, 179.

39. Ibid., 163.

40. Broadside advertisement for "Hagenbeck's Zoological Paradise," Hamburg, 1898. Translated from the French by Lydia Spitzer.

41. Ibid.

42. Ibid.

43. The fanciful qualities of the picture make it all the more unbelievable. Despite the advertisement's assertion that the scene was drawn from life, several factors, including the strange orientations of the animals and their often bizarre proportions, suggest that the artist took a good deal of liberty with the image.

44. Niemeyer, 64–66. See also Stephan Oettermann, *Das Panorama: Die Geschichte eines Massenmediums* (Frankfurt: Syndikat, 1980), 76.

45. For the official account of this story, see C. Hagenbeck, 31–32.

46. Wilhelm Munnecke, *Mit Hagenbeck im Dschungel* (Berlin: Scherl, 1931), 101.

47. Oettermann, 187–216.

48. As cited in ibid., 209.

49. As quoted in Dolf Sternberger, "A Panorama," in *Panorama of the Nineteenth Century,* trans. Joachim Neugroschel, intro. Erich Heller (New York: Urizen, 1977), 7.

50. As quoted in ibid., 8–9.

51. Ibid., 8.

52. Indeed, the concept behind his panoramas was different enough that he received Patent #91492 from the Imperial Patent Office for his "Panorama" in 1896 (Niemeyer, 64).

53. Quoted from the *Illustrierte Zeitung* by Niemeyer, 66.

54. Dr. Kurt Priemel, "Die Heutigen Aufgaben der Tiergärten. Eine Erwiderung," *Zoologischer Beobachter* 50 (1909): 365–66.

55. Dr. Kurt Priemel, "Handelstierpark und Zoologische Gärten," *Frankfurter Zeitung und Handelsblatt,* April 24, 1909.

56. Ibid.

57. Ibid.

58. Zukowsky, 167.

59. Ibid., 58.

60. Thinking of the view of the main panorama and its grazing-animal paddock, Alexander Sokolowsky seems to fall naturally into speaking about the park with the dual paradigms of Ark and Eden. He writes: "There the viewer sees, in trustful unity, zebras, eland, gnus, and many other creatures, peacefully moving about each other unconcerned about each other's activities. A practiced animal observer, however, will quickly notice that the different species keep to their own, just like in Noah's Ark, where the pairs were brought together by Father Noah, or in the pictures of the animal paradise of which we can thank the imagination of medieval artists." See *Carl Hagenbeck und sein Werk* (Leipzig: E. Haberland, 1928), 48.

61. Zukowsky, 61.

62. Sokolowsky, 43.

63. Zukowsky, 61–62.

64. The idea that Hagenbeck's was foremost a *Handelsmenagerie,* that is, an exotic animal dealership, was made clear by many authors, including Priemel, Heck, and Knauer.

65. The following discussion of animal prices is derived from price lists prepared by three traders: Carl Hagenbeck, Josef Menges of Limburg, and H. L. Hammerstein of Chicago. Old price lists, items that many zoos have in their collections, present a range of problems for historians. As will be seen in this discussion, the value assigned to an individual animal depended on a great many factors, and even though one list in 1890 advertised a lion for one price and another list in 1910 advertised another lion for a greater or lesser amount, it is impossible to conclude from this simple fact that the value of lions had somehow risen or declined by the difference in the two prices. Other problems in working with these mate-

rials stem from the multiple currencies involved in these transactions. One list will present animals in marks, another in pounds, and still another in dollars, and, of course, the strength of each of these currencies varied over time. For the purpose of this discussion, I have limited myself to price lists issued from 1899 to 1902, with the exception of Carl Hagenbeck's handwritten list of 1897, which is mentioned in the discussion of the albino "Thibetbear" in note 70 below, and, for most of the main discussion, I have used lists employing dollars. The lists used in this section can all be found in the box entitled "Animal Dealers Catalogs" in the Wildlife Conservation Society Archives, New York Zoological Park, Director's Office Records. The following lists were used: Carl Hagenbeck, handwritten list, "February 1897"; Carl Hagenbeck, typed list, "End of November 1899"; Carl Hagenbeck, typed list, ca. 1900; Carl Hagenbeck, undated printed list, listing also performance groups, ca. 1901–2; Carl Hagenbeck, printed list, "1 November 1901"; Josef Menges, handwritten list, "June 1900"; Josef Menges, printed list, "September 1901"; Josef Menges, printed list with handwritten addenda, "August 1902"; H. L. Hammerstein, handwritten list, "January 1st 1902"; and H. L. Hammerstein, typed list, "October 30th 1902."

66. "List of Animals Now for Sale—1 November 1901," Carl Hagenbeck's Handelsmenagerie und Thierpark, Hamburg, Wildlife Conservation Society Archives, New York Zoological Park, Director's Office Records.

67. Carl Hagenbeck to William T. Hornaday, March 2, 1910, Incoming Correspondence, Director's Office, New York Zoological Park, Wildlife Conservation Society Archives.

68. William T. Hornaday to Carl Hagenbeck, October 15, 1910, Outgoing Correspondence, Director's Office, New York Zoological Park, Wildlife Conservation Society Archives.

69. A price list from Hagenbeck dated November 1899, for example, highlighted: "Male Crossbreed of Lion & Tigress, born 11. March 1897, a very large animal & the only one in existence [*sic*]. He has a young tigress, 2 years old, & an imported Somali Lioness, 2 years old, as companions, & all three are very tame.—Lot: $4,000." See the "Animal Dealers Catalogs" in the Wildlife Conservation Society Archives, New York Zoological Park, Director's Office Records. Hagenbeck experimented a good deal with crossbreeding, partly in an effort to acclimatize certain species and partly because of the potential profits from the results of the breedings. Perhaps his most sustained efforts were directed to crosses between zebras and Arabian horses, from which he produced some quite striking, dark, dimly striped offspring.

70. Hagenbeck's list of 1897 shows again how animals could gain substantial value for being somehow unique. While the list prices "Malay bears" at £5 and a three-year-old female "Thibetbear" at £30, a five-year-old female "<u>albino</u>" "Thibetbear" was listed for £100.

71. "Carnivorous" and "Hay-Eating" were the typical divisions used by dealers in price lists of the period.

72. H. L. Hammerstein of Chicago, for example, listed in September 1902 "1 male

Giraffe, 1 year old, 8 feet high, in the most perfect condition imaginable, tame, $4000," delivered to Hoboken at shipper's cost and risk. On the other hand, Josef Menges of Limburg, Germany, offered a pair of giraffes, each more than eight feet in height, for $5,300 and another one-year-old male of the same height for $2,700 delivered "on board of the steamer, either in Hamburg, Bremen, Amsterdam or Antwerp." He also indicated, however: "Freight and perils of voyage, on charge of purchaser." He would undertake the task of bringing them to the United States, at his risk and expense, for $3,700 for the male and $7,300 for the pair. See H. L. Hammerstein, price list, September 30, 1902, and Josef Menges, handwritten price list, August 11, 1902, in "Animal Dealers Catalogs" in the Wildlife Conservation Society Archives, New York Zoological Park, Director's Office Records.

73. A 1903 bill from Carl Hagenbeck to the Bronx Zoo for animals to be housed in the new Antelope House, a single shipment with a value of close to $14,000, gives a sense of the relative costs of the various animals as well as a sense of Hagenbeck's ability to fulfill demanding orders. The bill itemizes:

1 pair Giraffes	$5,500.00
1 pair White Bearded Gnus	$1,500.00
1 pair White Tailed Gnus	$1,250.00
1 male Eland Antelope	$1,250.00
1 male Baker's Antelope	$800.00
1 male Beatrix Antelope	$725.00
1 female Addax	$600.00
1 pair Cervicapra Antelopes	
(female defective and to be replaced by another)	$200.00
1 female Cervicapra Antelope, with young male	$150.00
1 male Redunca Antelope	$150.00
1 male Isabelline Antelope	$125.00
2 female Duiker Antelope	$100.00
1 pair Llamas, Gift of Mr. Brewster	$400.00
1 pair Guanacos, Gift of Mr. Brewster	$400.00
1 male Vicuna, Gift of Mr. Brewster	$200.00
1 female Alpaca, Gift of Mr. Brewster	$200.00

William T. Hornaday to Carl Hagenbeck, October 27, 1903, Outgoing Correspondence, Director's Office, New York Zoological Park, Wildlife Conservation Society Archives. In addition to listing the animals in the bill, Hornaday writes, "I need hardly assure you that we are greatly pleased with these animals, and are fully appreciative of the fact that many of them are very rare, and some of them on the verge of extinction in the wild state. Certainly, there is no other man than yourself living who could have brought together such a collection and delivered it in fine condition at our door. But for your influence with the Duke of Bedford, we would of course have no eland, and in paying your bills, I also wish to extend the thanks of the Zoological Society, in general, and of its Director, in particular, for the attention you have given this order."

74. Purchasing the larger primates often turned into a quick financial loss for zoological gardens when, as was typical, the animals promptly died. In the fall of 1905, for example, the Bronx Zoo purchased the full-grown female gorilla "Miss Crowther" from the English dealer J. D. Hamlyn for £200, only to have the animal die, without insurance, during its voyage to the United States. Responding to the queries of the Bronx Zoo after the Hamlyn affair for further gorillas, the Hagenbecks suggested that they could purchase a gorilla in West Africa but would do so only if the zoo would agree to an arrangement about insurance, risk, and length of survival ahead of time. In the end, Hagenbeck's son Heinrich proposed in a letter to Hornaday: "I will try my best to insure the Gorillas but am afraid that H. B. Sedgewich and Co. will not be inclined to insure any more Gorillas. If the animal dies on the way over you stand ½ net loss of the animal, and if we have a chance of insuring it, you pay half after deducting the insurance. If it reaches the Zoo Park in good health, I am to get $500.— If it lives during the first week after its arrival, I am to get another $250.—at the end of this week. If it lives during the second week I am to get another $250.—at the end of this week, and another $250. at the end of the 3rd week. If it lives still 6 months later I get another $250.—I think this is a reasonable agreement and if you accept my offer, let me know. It was also understood from our conversation that nothing was agreed about a certain age. The younger the animals are, the better they will live, but it is understood that the animals must feed alone." See Heinrich Hagenbeck to William T. Hornaday, October 26, 1905, Incoming Correspondence, Director's Office, New York Zoological Park, Wildlife Conservation Society Archives. In this unusual arrangement, if the gorilla died in transit, the Bronx Zoo would be charged half of the original cost of obtaining the animal in West Africa. If, however, it survived for six months, Hagenbeck would receive a total of $1,500.

75. By the fall of 1911 Hagenbeck himself owned three different ostrich farms, one in Stellingen, one on the island of Brioni in the Adriatic, and one in German Southwest Africa, and he had plans for a fourth in German East Africa and a fifth in Berlin. With these, he wrote in a letter to William T. Hornaday of October 10, 1910, he "planned to produce enough feather to supply the German market" (Incoming Correspondence, Director's Office, New York Zoological Park, Wildlife Conservation Society Archives). Josef Menges became one the most consistent dealers in ostriches, which he usually captured in the Sudan. His correspondence with ostrich farms gives some idea of this growing market. In December 1899, for example, he received the following from Morton Taylor, president of "The Florida Ostrich Farm," "Ostrich Breeders and Manufacturers of all kinds of Ostrich goods":

Mr. Edward Atherton, who is the Supt. on our Ostrich Farm at Fullerton, California, has forwarded us your letter of Nov. 16th, in which you state you have some of the North African young ostriches for sale. We have thought seriously of interbreeding our S. African birds with the N. African birds, and regrat [*sic*] you have no older birds for sale the present time.

We could no doubt buy your birds in a few weeks time, but just at the present moment, we are rather upset, concentrating our different farms, pairing off our birds, building fences, etc. We would thank you to advise us in regard to duty, if there is any, between Germany and the United States on young birds, and that if The Florida Ostrich Farm Inc. bought this lot of twenty birds, from you, if you could supply us with more, were we to give you a reasonable time to capture them, and if it is possible to secure any older ones. If you will kindly advise us in these different matters, we will then give you our answer as to our needs, as well as to the lot of twenty birds, of which we have already spoken.

The Florida Ostrich Farm, in addition to the farm at this place, owns the Fullerton Farm, and the Arizona Farm, consisting in all, about six hundred ostriches, and we should be glad to experiment by interbreeding as before stated.

See Correspondence of Josef Menges, Gift of the Menges family, Hagenbeck Archive, Stellingen.

76. Hagenbeck's November 1899 price list includes a reticulate python, 22–24 feet in length, at $400; Menges in September 1901 lists another reticulate python of 8 meters for 1,200 reichsmarks. See "Animal Dealers Catalogs" in the Wildlife Conservation Society Archives, New York Zoological Park, Director's Office Records.

77. In fact, in one catalogue eleven different performing groups are listed for sale. "Group 2," for example, consisted of: "1 male Lion 'Pascha,' 3½ years; 1 male Lion 'Prince,' 3½ years; 1 male Lion 'George,' 2¾ years; 1 male Lion 'August,' 5 years, 1 eye blind; 1 male Chinese Tiger 'Willy,' 3 years; 1 male Sumatra Tiger 'Adam,' 3½ years; 1 female Sumatra Tiger 'Eva,' 3½ years; 1 male Leopard, 'Sam,' 3½ years; 1 male Polarbear 'Muffel,' 3 years; 1 male Thibetbear 'Asia,' 4 years; 4 German Boarhounds; All properties, 1 Central cage, 2 new Caravans, included—the lot £2750." Carl Hagenbeck, undated printed price list, ca. 1900–1901. "Animal Dealers Catalogs" in the Wildlife Conservation Society Archives, New York Zoological Park, Director's Office Records.

78. *Perfect* is a term used by dealers to describe animals without faults which conform to expectations for the species.

79. William T. Hornaday to Carl Hagenbeck, November 20, 1912, Outgoing Correspondence, Director's Office, New York Zoological Park, Wildlife Conservation Society Archives. Hagenbeck related the asking price in a letter of November 9, 1912.

80. Why pygmy hippos and why so much? To answer these questions, one needs to look at the history of the species and the remarkable story of their capture. Although the pygmy hippopotamus was first described in 1844, and although skeletons and skins of the animal had been deposited in a handful of natural history museums, only two of the creatures had made it to Europe by the end of the nineteenth century—one lived for a few hours at the Dublin Zoo in 1873, and Hagenbeck had one at his menagerie in 1885. By 1912 the

pygmy hippopotamus was still essentially an entirely unknown, almost mythical creature. As Hornaday noted in the *Bulletin* of the New York Zoological Society: "With the exception of a few museum men, and the few zoologists who are specially interested in the ungulates, the Pygmy Hippopotamus has been to the world nothing more than a name, and to most people it has been not even that" (William T. Hornaday, "Our Pygmy Hippopotami," *Zoological Society Bulletin* 16, no. 52 [1912]: 877). Noting that there were always "bold and venturesome men," however, who were willing to go after the rarest animals for a price, Hornaday recalled in his brief history of the Bronx Zoo's hippos that Hagenbeck, "ever ready to try the untried, and attempt the impossible," dispatched an "intrepid hunter and explorer" to catch these rarest creatures (878). Hagenbeck's correspondence with Hornaday suggests that he felt the animals must be nearing extinction and that the large investment he had made in catching the animals, which he estimated would exceed $15,000 for the five initial animals, was worth the effort to get a few of the pygmy hippos alive (Carl Hagenbeck to William T. Hornaday, January 26, 1912, Incoming Correspondence, Director's Office, New York Zoological Park, Wildlife Conservation Society Archives). He explained to Hornaday that he had emphasized to his hunter that he should not shoot any of the animals until at least two or three specimens were safely in captivity, a point that Schomburgk himself recalls in his story of catching the hippos ("On the Trail of the Pygmy Hippo: An Account of the Hagenbeck Expedition to Liberia," *Zoological Society Bulletin* 16, no. 52 [1912]: 880).

In the end, it took Schomburgk two expeditions into the interior of Liberia over the course of some eighteen months to catch the hippos for Hagenbeck, and Hornaday quickly purchased two and then a third of the animals. The announcement of their imminent arrival in New York was well staged by Hagenbeck and Hornaday. Photographs and an account of the Schomburgk's adventures among the "cannibals" of Liberia were prominently featured in press releases from the zoo, and Schomburgk's account of his perils in catching the animals and controlling his somehow always-ready-to-rebel African workers and carriers was quickly printed in the *Bulletin* of the zoological society and in the *New York Times* (July 14, 1912, sec. 5, p. 14; see also William T. Hornaday to Carl Hagenbeck, July 25, 1912, Outgoing Correspondence, Director's Office, New York Zoological Park, Wildlife Conservation Society Archives). According to Hornaday's letters to Hagenbeck, several gentlemen of the society strongly objected to the high cost for the two animals, but Hornaday was apparently convinced that the hippos would be among the most remarkable animals in his collection, and clearly the idea that he would have the only pair in any zoological garden constituted a large part of his interest in them.

81. Christoph Schulz, *Auf Großtierfang für Hagenbeck: Selbsterlebtes aus afrikanischer Wildnis* (Dresden: Deutsche Buchwerkstätten, 1921), 184.

82. For an entrepreneur who was very interested in promoting the construction of new zoological gardens, some of his reasons for setting an inexpensive example are obvious.

83. Niemeyer, 64.

84. Rosl Kirchshofer, "Environmental Education in Zoological Gardens," *Garten und Landschaft* 1 (1985): 32–33.

85. Heini Hediger, *Man and Animal in the Zoo: Zoo Biology,* trans. Gywnne Vevers and Winwood Reade (New York: Delacorte, 1969), 188.

86. Hagenbeck's ideas were quickly taken up by other zoos. First of all, older zoos quickly constructed individual exhibits based on Hagenbeck principles. In 1914, for example, P. C. Mitchell, the secretary of the London Zoological Society, enthusiastically called Stellingen "frankly theatrical scenery," scenery that, moreover, served as the principal inspiration for his and J. P. Joass's designs for the Mappin Terraces at the gardens of the same year. See Sir Peter Chalmers Mitchell, *Centenary History of the Zoological Society of London* (London: Zoological Society of London, 1929). Eventually, whole zoos, such as those in Rome, Vincennes, Munich, and Detroit, followed Hagenbeck plans.

87. Zukowsky, 59–60.

88. Ibid., 60.

89. C. Hagenbeck, 414–15. It is also worth noting that Hagenbeck wrote that five ships took part in the kill. Zukowsky—perhaps for greater effect—has reduced that number to one.

90. Sokolowsky, 174.

91. Ibid., 175.

92. Priemel, "Handelstierpark" (original emphasis).

93. Alexander Sokolowsky, "Kleine Hagenbeck-Erinnerungen," *Der Zoologische Garten* 21, no. 1 (1954): 10–11.

94. Ibid., 11.

95. Karl Max Schneider, "Zuneigung," *Der Zoologische Garten* 21, no. 1 (1954): 1.

Conclusion *When Animals Speak*

1. See Robert Yerkes, *Almost Human* (London: Cape, 1925).

2. For more on Johnson, see James W. Cook Jr., "Of Men, Missing Links, and Nondescripts: The Strange Career of P. T. Barnum's 'What Is It?' Exhibition," *Freakery: Cultural Spectacles of the Extraordinary Body,* ed. Rosemarie Garland Thomson (New York: New York University Press, 1996), 139–57.

3. In the Archives of the New York Zoological Society, there are several clippings about "Miss Crowther" collected from London newspapers by Cecil French, the dealer who organized the purchase of Crowther from J. D. Hamlyn for the Bronx Zoo. Among the clippings is one from the *Star* with the title "A Congo Beauty: The Zoo's New Gorilla Talks of First Impressions," presumably from late September 1905. Correspondence between Hamlyn and French, French and Hornaday, and Hagenbeck and Hornaday regarding Crowther is also in the collection. On the gorilla John Daniel, see William T. Hornaday, "A Gorilla That Lived Like a Human," *Mentor* (November 1921): 30–31; "John Daniel, Civilized

Gorilla," *Literary Digest* December 10, 1921: 44–48; "Circus's Gorilla a Bit Homesick," *New York Times* April 3, 1921; L. B. Yates, "Gambling in Jungle Stuff," *Saturday Evening Post* January 20, 1923, 16 ff.; and Emily Hahn, "Annals of Zoology: A Moody Giant—I, "*New Yorker* August 9, 1982: 39–61. For a description of the arrival of John Daniel II in 1924, see "Circus to Greet John Daniel 2d," *World* April 6, 1924, 14, and "John Daniel 2d, $100,000 Gorilla, to Arrive To-Day," *World* April 6, 1924, 12E. My sincere thanks to Richard Reynolds III for sharing his remarkable knowledge of the history of the exhibition of nonhuman primates.

4. Franz Kafka, "A Report to an Academy," trans. Willa and Edwin Muir, in *Franz Kafka: The Complete Stories,* ed. Nahum N. Glatzer (New York: Schocken, 1971), 250.

5. Ibid., 257.

6. Rainer Maria Rilke, "The Panther, *Jardin des Plantes, Paris,*" in *Neue Gedichte,* trans. Stephen Cohn (Manchester: Carcanet, 1992).

7. Alexander Sokolowsky, *Carl Hagenbeck und sein Werk* (Leipzig: E. Haberland, 1928), 170–71.

8. For this quotation, I have used the English translation of Hagenbeck's memoir, *Beasts and Men: Being Carl Hagenbeck's Experiences for Half a Century among Wild Animals* (London: Longmans, 1912), 291–92. The quotation is a direct translation from the unabridged 1908 German edition, *Von Tieren und Menschen: Erlebnisse und Erfahrungen* (Leipzig: Paul List, 1908), 436–37.

9. The caption in the early English editions reads simply, "The Three Friends."

10. My thanks to Marcus Bullock for his help in clarifying the sense of the caption.

11. I wish to thank the members of the Center for Twentieth Century Studies 1997 research group, in particular Jeffrey Hayes, Theresa Mangum, and Lane Hall, along with Heather Hathaway, for their thoughtful responses to the picture.

12. Peter Guillery, *The Buildings of London Zoo* (London: Royal Commission on the Historical Monuments of England, 1993), 43.

13. For a particularly strong essay on "immersion displays," see Jeffrey Hyson, "Jungles of Eden: The Design of American Zoos," in *Environmentalism in Landscape Architecture,* ed. Michel Conan (Washington, D.C.: Dumbarton Oaks, 2000), 23–44.

14. Vicki Croke, *The Modern Ark: The Story of Zoos: Past, Present, and Future* (New York: Avon, 1997), 95.

15. See C. Hagenbeck, *Von Tieren und Menschen,* 377–86.

16. Günter H. W. Niemeyer, *Hagenbeck: Geschichte und Geschichten* (Hamburg: Christian, 1972), 214.

17. Ibid.

18. The fish that one sees in large public aquariums and in private saltwater tanks in homes have almost always been caught in the wild. In a 1989 essay on the pet trade in marine fish, Rolf Möltgen notes that while saltwater fish are caught off the Caribbean Islands, Hawaii, East Africa, Ceylon, the Maldives, Indonesia, and the Philippines, most come from

the Philippines, where in some months more than 200,000 fish worth 600,000 deutsche marks are caught *each month* using poisons put into the water. Less than half the fish survive to their destination. See "Das Krokodil in der Badewanne: Über den alltäglichen Umgang mit exotischen Tieren," in *Wir töten, was wir lieben: Das Geschäft mit geschützten Tieren und Pflanzen,* ed. Dieter Kaiser (Hamburg: Hoffmann und Campe, 1989). Of course the trade in wild-caught freshwater tropical fish can also be surprisingly large. A 2001 estimate by scientists and aquarists at the New England Aquarium and the Universidade do Amazonas claims that some 20 million live fish are exported annually from the Brazilian Rio Negro basin and that most of these (in some areas, as many as 80 percent) are the small red and blue cardinal tetras. In the case of the cardinal tetra fishery, however, the authors claim that the fishery is a model sustainable industry that actually protects the forest from more destructive forestry, agriculture, and mining. Indeed, the authors worry that the farming of cardinal tetras in the United States might threaten the largely commendable tetra trade stemming from South America. See Ning Labbish Chao, Scott Dowd, and Michael Tlusty, "Project Piaba: Buy a Fish, Save a Tree," *Communiqué* (January 2001): 14–16. Meanwhile, in another South American example, Timothy Wright et al. have reported that as many as 800,000 parrots are taken annually from nests in Central and South America for the black-market pet trade. Noting that from 30 to 75 percent of young are being taken from the nests, the authors worry that many species of parrots may see a collapse in twenty years as the current adults die off. See T. F. Wright et al., "Nest Poaching in Neotropical Parrots," *Conservation Biology* 15 (2001):710–20.

A Note on Sources

In the library of the Milwaukee County Zoo, amid shelves and cabinets filled with scientific journals and the publications of other zoological gardens, there is one shelf with a few older-looking books. One of the volumes is Elisabeth Schulz's *Afrikanische Nächte: Erzäh-lung* (Hamburg: Zoo Verlag [Christoph Schulz], 1926), her semifictional account of her adventures catching animals with her husband, Christoph, in German East Africa before World War I. There are probably no more than a dozen copies of this little volume in American libraries, and here was an autographed one a couple of miles from my home. Over the years I have become accustomed to finding pieces from the history of Carl Hagenbeck and his firm in the most unexpected places. In fact, I am fairly certain that it was reading the quick reference to Hagenbeck in Franz Kafka's "A Report to an Academy," trans. Willa and Edwin Muir, in *Franz Kafka: The Complete Stories,* ed. Nahum N. Glatzer (New York: Schocken, 1971), 250–62, which probably began this whole project. Along the way there have been some truly memorable finds. Among the more rewarding were spending a weekend with the daughter-in-law of Christoph and Elisabeth Schulz at her son's ranch in Texas, where we pored over albums of photographs from the early years of the twentieth century; visiting the library of the Wildlife Conservation Society in New York (formerly the New York Zoological Society), where I found an extensive collection of letters between William Temple Hornaday and Carl Hagenbeck, as well as a well-informed archivist in Steve Johnson; corresponding with Jutta Niemann, the granddaughter of Hans Hermann Schomburgk, who was in the midst of preparing what became a beautiful museum exhibit about the famous man whom she admired so greatly; meeting Ken Harck, whose extensive collection of circus memorabilia included a stunning mint-condition poster of the Hagenbeck exhibit at the Chicago Columbian Exposition of 1893; and, of course, visiting the archives of Hagenbeck's Tierpark in Stellingen.

With a company like Carl Hagenbeck's, one can find archival materials all over the world. A stunning variety of people came into contact with Hagenbeck, and a great many retained correspondence and other materials while recording as well their thoughts in personal papers. A wonderful portrait of Carl Hagenbeck shows him seated at his desk in the early years of the twentieth century, taking care of his correspondence, while a picture of his father, framed in ivory boars' tusks, looks on. Hagenbeck's firm was largely based in daily business correspondence, and just about any person or institution owning a substantial exotic animal collection at the end of the nineteenth century is likely to have materials of some scope relating to the firm. I have not been able to look everywhere, however, and for the benefit of future researchers, I would like especially to encourage research into the papers of private animal buyers such as the duke of Bedford, Baron Walter von Rothschild, the Russian Frederic von Falz-Fein, and others. Further, records of the major zoological gardens and circuses of the latter half of the nineteenth century might well hold gems of Hagenbeck history. I draw a close to my research for this book knowing that there is much yet to be discovered.

The contemporary materials on which this study is primarily based represent the most expansive work done on the firm of Carl Hagenbeck so far. They are generally of two kinds: manuscript, photographic, pamphlet, and other records in public and private archives and collections; and published accounts of the firm and its activities in popular or scientific journals, magazines, and newspapers, as well as in memoirs, biographies, and other such materials. The primary archival sources consulted include materials from the Hamburgisches Museum für Völkerkunde, the library of Wildlife Conservation Society in New York, and the Hagenbeck Archive in Stellingen. In addition, many other private and public collections have been consulted, including those of the Chicago Historical Society, the Schulz family in Texas, the Circus World Museum in Wisconsin, and the Smithsonian collections in Washington, D.C.

The most important archival source for materials on Carl Hagenbeck is probably that of the firm itself. Though heavily damaged during World War II, the archive contains an extensive collection of materials salvaged by the care and concern of the longtime Hagenbeck associate Günter Niemeyer. Included in the collection are a variety of documents, photographic albums and individual photographs carefully catalogued, various publications of the firm, books of clippings, salvaged letters, and a largely complete set of the firm's account books going back to the last decades of the nineteenth century. One of the strongest aspects of the archive is its extensive photographic collection, which is well documented and serves as the base for occasional exhibitions on the firm's history. Because the archive is privately owned, however, access to it is limited and dependent on permission from firm representatives. In 1998, Lothar Dittrich and Annelore Rieke-Müller published, with the support of the firm, an anniversary biography entitled *Carl Hagenbeck (1844–1913): Tierhandel und Schaustellungen im Deutschen Kaiserreich* (Frankfurt: Peter Lang, 1998), which is the first full-length account based primarily on the archival materials of the firm. While focusing on what the authors contend is "the most important lasting contribution of Carl Hagenbeck, namely on the reception of his animal enclosures in Stellingen" (12), the book is a major contribution to knowledge of the company, and the authors and the firm are to be commended for their work. Praise, especially for its photographs, is due as well to another anniversary book, the coffee-table-destined *Hagenbeck: Tiere, Menschen, Illusionen* (Hamburg: Hamburger Abendblatt, 1998), by Matthias Gretzchel and Ortwin Pelc, completed again with support from the firm along with the Museum für Hamburgishe Geschichte, which hosted a 150th anniversary exhibition.

The library of Hamburgisches Museum für Völkerkunde holds the papers of Johan

Adrian Jacobsen, who organized his first "people show" for Carl Hagenbeck in 1878 and who remained closely affiliated with the firm until his death. Jacobsen's papers consist of around five thousand letters and drafts; a series of diaries he kept on his travels, which were often intended to serve as the basis for later publications; and a great deal of assorted information about the exhibitions, including programs from his own and others' shows, clippings, and the like. For example, a number of the pamphlets referred to in the chapters on the animal trade and people shows, including *Carl Hagenbeck's Thier-Karawane aus Nubien* (Hamburg: n.p., n.d.), *J. Menges' Ost-Afrikanischen Karawane aus den Somalilande* (Hamburg: n.p., n.d.), and *Gebr.-Hagenbeck's Indische Carawane* (Hamburg: n.p., n.d.), are included in this collection. A large proportion of this material stems from Jacobsen's private correspondence, including letters from the various associates of the Hagenbeck firm, among the more interesting of which are letters from Hagenbeck himself and one of his more important animal catchers, and eventually close associate of Jacobsen, Josef Menges. Hilke Thode-Arora based her *Für fünfzig Pfennig um die Welt: Die Hagenbeckschen Völkerschauen* (Frankfurt: Campus, 1989) on the Jacobsen papers. Her careful and thoughtful book has been the only full-length study of the firm to be prepared by someone not directly affiliated with or supported by the company.

Among the archival sources in the United States, the most important for this study have been the collections of the Wildlife Conservation Society in New York (formerly the Bronx Zoo), for material about the animal trade and exhibition, and the Chicago Historical Society, for materials relating to the 1893 World's Columbian Exposition, for which Hagenbeck constructed a large pavilion on the Midway Plaisance. Indeed, perhaps the most important single collection of correspondence relating to Hagenbeck is to be found in the library of the Wildlife Conservation Society. Uniquely, the library preserves both Hagenbeck's incoming letters and copies of William Temple Hornaday's responses. Both the incoming and outgoing correspondence to the Bronx Zoo are in the process of being filmed for microfiche and will be an important resource for scholars interested in the animal trade for many years to come. Beyond these collections, scholars interested in the history of circuses should consult the collections of the World Circus Museum in Baraboo, Wisconsin, as well as the articles appearing in *Bandwagon,* the journal of the Circus Historical Society, edited by Fred Pfening Jr. Finally, Jurgen Schulz allowed me to access his large collection of materials relating to the animal trade in the first decades of the twentieth century and to his parents, Walter and Ursula Schulz, and his grandparents, Christoph and Elisabeth Schulz. This is a one-of-a-kind collection of materials detailing the lives of a German family that moved to German East Africa in the first decade of the twentieth century to become animal catchers. A century later, the family is still involved in the exotic animal business.

The contemporary published accounts of Carl Hagenbeck and his firm which have been used in this study can be conveniently divided into four sections: (1) accounts of the firm in general, (2) accounts of the adventures of the Hagenbeck animal catchers, (3) accounts of the Hagenbeck people shows, and (4) accounts of the Hagenbeck performing animal acts and the Animal Park in Stellingen.

The most important contemporary account of Carl Hagenbeck and his company remains Hagenbeck's autobiography, *Von Tieren und Menschen: Erlebnisse und Erfahrungen* (Leipzig: Paul List, 1908), prepared with the help of Philipp Berges. The book appeared in numerous editions and translations and both encapsulated earlier accounts and helped mold a new vision of the company as being primarily rooted in a fascination and respect for the world's animal and human life. Earlier full-length biographies included the particularly important *Lebensbeschreibung des Thierhändlers Carl Hagenbeck,* by Heinrich Leutemann (Hamburg: Carl Hagenbeck, 1887), and Wilhelm Fischer's *Aus dem Leben und Wirken eines interessanten Mannes* (Hamburg: Baumann, 1896). Later but nevertheless contemporary accounts include Alexander Sokolowsky's uncritical *Carl Hagenbeck und sein Werk* (Leipzig: E. Haberland, 1928) and Ludwig Zukowsky's similar work, *Carl Hagenbecks Reich: Ein deutsches Tierparadies* (Berlin: Wegweiser, 1929).

Among the people who worked as animal catchers for Hagenbeck, many left accounts of their adventures. Lorenzo Casanova, who was the first professional catcher for Hagenbeck, is discussed in varying detail in practically all the accounts of Carl Hagenbeck, including Hagenbeck's own memoirs and the biography of Leutemann. In addition, H. Dorner wrote an article entitled "Casanova und Hagenbeck" for *Die Gartenlaube* 69 (1869): 42–47, which includes an illustration by Leutemann entitled "Ankunft der Thierkarawane Casanova" (44). Other important early accounts include the pamphlets *Carl Hagenbeck's Thier-Karawane aus Nubien,* presumably about a shipment organized by Bernhard Cohn, and *J. Menges' Ost-Afrikanischen Karawane aus den Somalilande,* both included in the Jacobsen papers. Additional information about Menges can be found in a small collection of letters donated by the Menges family at the Hagenbeck Archive, as well as in a series of articles Menges prepared for the geographical journal *Petermanns Mittheilungen.* The articles include "Jagdzug nach dem Mareb und oberen Chor Baraka, März und April, 1881," *Petermanns Mittheilungen* (1884): 162–69; "Ausflug in das Somaliland," *Petermanns Mittheilungen* (1884): 401–10; "Reisen zwischen Kassala und dem Setit," *Petermanns Mittheilungen* (1888): 65–67; "Die Karawanenstrassen zwischen Suakin und Kassala," *Petermanns Mittheilungen* (1887): 97–101; and "Das unbekannte Horn von Afrika," *Petermanns Mittheilungen* (1889): 49–51. Menges also prepared a scientific article on the Somali wild ass, "Der Wildesel des Somalilandes. (*Equus asinus somalicus.*)," *Der Zoologische Garten* 28, no. 9 (1887): 261–68.

Carl's half brother John Hagenbeck, who lived for most of his adult life in Sri Lanka, where he managed several plantations and regularly shipped animals and people shows to Europe, published several volumes of reminiscences including *John Hagenbecks abenteuerliche Flucht aus Ceylon: Meine Ausweisung aus Ceylon und Flucht nach Europa* (Dresden: Deutsche Buchwerkstätten, 1917); *Fünfundzwanzig Jahre Ceylon: Erlebnisse und Abenteuer im Tropenparadies,* ed. Victor Ottmann (Dresden: Deutsche Buchwerkstätten,

1922); *Kreuz und Quer durch die Indische Welt: Erlebnisse und Abenteuer in Vorder- und Hinterindien, Sumatra, Java, und auf den Andamanen,* ed. Victor Ottmann (Dresden: Deutsche Buchwerkstätten, 1922), and *Südasiatische Fahrten und Abenteuer: Erlebnisse in Britisch- und Holländisch-Indien, Im Himalaya und in Siam,* ed. Victor Ottmann (Dresden: Deutsche Buchwerkstätten, 1924). Wilhelm Munnecke published a partly fictionalized account of John Hagenbeck entitled *Mit Hagenbeck im Dschungel* (Berlin: Scherl, 1931). John Hagenbeck's adventures during his many years in Sri Lanka and India offer an important record, and his observations of the indigenous people, whether Tamil, Sinhalese, or Vedda, are clear and concise and rarely burdened with ignorance.

Two Hagenbeck hunters of the first decades of the twentieth century who deserve special attention are Hans Hermann Schomburgk and Christoph Schulz. Schomburgk's many books detailing his careers first as an elephant hunter, then as an animal catcher, then as an explorer of Africa, and then finally as a photographer of African wildlife included *Wild und Wilde im Herzen Afrikas* (Berlin: Fleischel, 1910); *Bwakukama: Fahrten und Forschungen mit Büchse und Film im Unbekannten Afrika* (Berlin: Deutsch-Literarisches Institut, 1922); *Von Mensch und Tier und Etwas von Mir* (Berlin: H. Wigankow, 1947); *Der Kleine Jumbo, das Schicksal eines Afrikanischen Elefantenbabys* (Hannover: A. Weichert, 1952); and *Zelte in Afrika. Fahrten-Forschungen-Abenteuer in sechs Jahrzehnten* (Berlin: Verlag der Nation, 1960). Of all his books, *Wild und Wilde* is his most interesting and informative. Christoph Schulz began working for Hagenbeck in 1909 and had a long association with the firm. Beyond his recollections about hunting for Hagenbeck, Schulz's books would be useful to those interested in German settlement and farming in the former colony of German East Africa. Schulz's works include *Auf Großtierfang für Hagenbeck: Selbsterlebtes aus afrikanischer Wildnis* (Dresden: Verlag Deutsche Buchwerkstätten, 1921) and its slightly abridged version, *Aus Hagenbecks Jagdgründen: Abenteuer eines Tierfängers in den Steppen und Urwäldern Afrikas* (Leipzig: Verlag Deutsche Buchwerkstätten, 1922). His *Jagd- und Filmabenteuer in Afrika: Streifzüge in das Innere des dunklen Erdteils* (Dresden: Verlag Deutsche Buchwerkstätten, 1931) recounts the making in 1913 of a film to be shown at the theater at Hagenbeck's Tierpark. Finally, as I noted above, Schulz's wife, Elisabeth Schulz, published her own account of their adventures in Africa titled *Afrikanische Nächte: Erzählung.*

Among briefer accounts of catching for Hagenbeck are chapters in Hans Dominik's *Kamerun: Sechs Kriegs- und Friedensjahre in deutschen Tropen* (Berlin: Mittler, 1901) and *Vom Atlantik zur Tschadsee: Kriegs- und Forschungsfahrten in Kamerun* (Berlin: Mittler, 1908); C. G. Schillings' *Flashlights in the Jungle,* trans. Frederic Whyte (New York: Doubleday, 1905), and *Der Zauber des Elelescho* (Leipzig: R. Voigtländer, 1906); Ludwig Heck's "Die Schillingsche Sendung Deutsch Ostafrikanische Tiere," *Die Gartenlaube* (1900): 562–64; an account of Dietrich Hagenbeck entitled "Ein Junger Nilpferdjäger: Dietrich Hagenbeck," *Die Gartenlaube* (1873): 754; and Hermann Wiele's *Für Hagenbeck im Himalaja und den Urwäldern Indiens: 30 Jahre Forscher und Jäger* (Dresden: Verlag Deutsche Buchwerkstätten, 1925).

Perhaps the most important contemporary source for the exhibitions of people are the many lectures delivered by Rudolf Virchow printed in the *Verhandlungen der Berliner Gesellschaft für Anthropologie, Ethnologie und Urgeschichte*. I will not list them all here, but among the more important for this study are "Zur Vorstellung der nach Berlin gebrachten Lappen," *Verhandlungen* 7 (1875): 225–28; "Über Eskimos," *Verhandlungen* 10 (1878): 57–58; "Über die in Berlin anwesenden Nubier," *Verhandlungen* 10 (1878): 333–55; "Über die ethnologischen Verhältnisse der Nubier," *Verhandlungen* 10 (1878): 387–407; "Über die Eskimos von Labrador," *Verhandlungen* 12 (1880): 253–74; "Über die Feuerländer," *Verhandlungen* 12 (1881): 375–393; "Über die Australier von Queensland," *Verhandlungen* 16 (1884): 407–18; "Über die Sogenannten Amazonen und Krieger Des Königs von Dahome," *Verhandlungen* 23 (1891): 114; "Über die Sogenannten Azteken und die Chua," *Verhandlungen* 23 (1891): 370–77; "Vorstellung von Lappen," *Verhandlungen* 23 (1891): 478–80; "Über die Dinka," *Verhandlungen* 27 (1895): 148–68; and "Über die beiden Azteken," *Verhandlungen* 33 (1901): 348–50. Similar scientific articles were prepared by other authors, such as Franz Boas's "Mittheilungen über die Lilxula-Indianer," *Original-Mittheilungen aus der Ethnologischen Abtheilung der Königlichen Museen* (Berlin) 1 (1886): 177–82.

Other significant contemporary accounts of the people exhibits include L. Beckmann, "Die Hagenbeck'schen Singhlesen," *Die Gartenlaube* (1884): 564–66; Heinrich Leutemann, "Lappländer, Nordische Gäste," *Die Gartenlaube* (1875): 740–44, and his "Die Eskimos in Paris," *Die Natur. Zeitung zur Verbreitung naturwissenschaftlicher Kenntnis und Naturschauung für alle Stände*, n.s., 4 (1878): 115–19; Heinrich Steinitz, "Die Feuerländer," *Die Gartenlaube* (1881): 732–35; Gustav Sundblad, "Samojeden im Zoologischen Garten zu Leipzig," *Die Gartenlaube* (1883): 95–97; and K. Boeck, "Sinhalesen Teufelstänzer," *Die Gartenlaube* (1901): 88. In addition, I made extensive use of two daily newspapers, the conservative *Die Neue Preußische (Kreuz-) Zeitung* (Berlin) and the *Frankfurter Zeitung und Handelsblatt* (Frankfurt), both of which reported at length on the shows. Additional work in daily papers from other cities should yield substantially more information. Among the more important accounts by Johan Adrian Jacobsen of his trips collecting people and artifacts for shows and ethnological museums are *Captain Jacobsen's Reise an der Nordwest-Küste Amerikas zum Zwecke Ethnolog. Sammlungen*, ed. A Woldt (Leipzig: M. Spohr, 1884); *Reise in die Inselwelt des Banda-Meeres*, ed. Paul Roland, intro. Rudolf Virchow (Berlin: Mitscher and Röstell, 1896); and *Alaskan Voyage, 1881–1883: An Expedition to the Northwest Coast of America* (Chicago: University of Chicago Press, 1977). Jacobsen also published many articles, including "A. Jacobsen's und H. Kühn's Reise in Niederländisch-Indien," *Globus. Illustrirte Zeitschrift für Länder- und Völkerkunde* 55, no. 11 (1889): 161–68; 12: 182–86; 13: 200–204; 14: 213–17; 15: 225–29; 16: 244–48; 17: 261–65; 18: 279–80; 19: 299–302; and "Leben und Treiben der Eskimo," *Das Ausland. Wochenschrift für Erd- und Völkerkunde* 64, no. 30 (1891): 593–98; 32: 636–39; 33: 656–58. For a more complete list of Jacobsen's publications, see Thode-Arora. Typical of the many books variously based on the photographic legacy of the shows are such works as Johannes Ranke, *Der Mensch* (Leipzig: Bibliographisches Institut, 1887); Hermann Heinrich Ploss and Max Bartels, *Woman: An Historical, Gynaecological, and Anthropological Compendium* (1913; London: Heinemann, 1935); Alexander Sokolowsky, *Menschenkunde: Eine Naturgeschichte sämtlicher Völkerrassen der Erde. Ein Handbuch für Jedermann* (Stuttgart: Union Deutsche Verlagsgesellschaft, 1901); and Carl Stratz, *Die Rassenschönheit des Weibes* (Stuttgart: F. Enke, 1902). Finally, perhaps the most enigmatic contemporary work about the function of people shows in European culture is Peter Altenberg's *Ashantee* (Berlin: Fischer, 1897).

For contemporary accounts of Hagenbeck's experiments in training animals and of his Animal Park, the best sources remain Hagenbeck's own *Von Tieren und Menschen,* along with Zukowsky's *Carl Hagenbecks Reich* and Sokolowsky's *Carl Hagenbeck und sein Werk.* The collections of the Circus World Museum and the Chicago Historical Society have perhaps the best materials available about Hagenbeck's animal training methods of the 1880s and 1890s. Among the better pieces I found there were Cleaveland Moffett, "Wild Beasts and Their Keepers: How the Animals in a Menagerie Are Tamed, Trained, and Cared For," *McClure's Magazine* (May 1894): 544–61; Cleaveland Moffett, "Wild Beasts in Captivity: How They Are Watered, Fed; the Special Dangers in Handling Them," *McClure's Magazine* (June 1894): 71–88; Raymond Blathwayt, "Wild Beasts: How They Are Transported and Trained," *McClure's Magazine* (July 1893): 126–35; Raymond Blathwayt, "The Training of Wild Beasts," *Pearson's Magazine* (1895): 240–46; *Hagenbeck's Arena and World's Museum,* official souvenir program, World's Columbian Exposition, Chicago Historical Society; *The World's Fair, Being a Pictorial History of the Columbian Exposition* (Chicago: n.p, 1893); a photographic album, *Midway Types* (Chicago: n.p., n.d.), Chicago Historical Society; and *Die Illustrirte Welt-Ausstellung, Chicago, 1893. Das Columbische Weltausstellungs-Album* (Chicago: Rand, McNally, 1893). Particularly useful pieces on the Animal Park before World War I include Friedrich Knauer's history *Der Zoologische Garten: Entwicklungsgang, Anlage und Betrieb unserer Tiergärten* (Leipzig: Theod. Thomas Verlag, n.d. [ca. 1911]); Kurt Priemel's "Handelstierpark und Zoologische Gärten," *Frankfurter Zeitung und Handelsblatt* (Frankfurt), April 24, 1909; and many articles in the journal of the Association of German Zoological Gardens, *Zoologischer Beobachter,* including E. E. Leonhardt, "Die Heutigen Aufgaben der Tiergärten," *Zoologischer Beobachter* 50 (1909): 321–28; Kurt Priemel, "Die Heutigen Aufgaben der Tiergärten: Eine Erwiderung," *Zoologischer Beobachter* 50 (1909): 354–66; Friedrich Katt, "Hagenbecks Tierparadies," *Zoologischer Beobachter* 50 (1909): 370–72; Ludwig Zukowsky, "Ueber einige seltene und kostbare Tiere im Carl Hagenbecks Tierpark," *Zoologischer Beobachter* 55 (1914): 179–87, 213–17, 228–34; "Schlangenkämpfen in Carl Hagenbecks Tierpark in Stellingen," *Zoologischer Beobachter* 54 (1913): 7–8; "Die Leoparden in Carl Hagenbecks Tierpark in Stellingen," *Zoologischer Beobachter* 54 (1913): 49–50; "Hagenbecks Zwergflusspferd," *Zoologischer Beobachter* 53 (1912): 132–34, 208–10, 338–39; Oscar de Beaux, "See-Elefanten in

Carl Hagenbecks Tierpark," *Zoologischer Beobachter* 52 (1911): 73–79; and "Strauss und Rhinozeros in Hagenbeckschen Tiergarten," *Zoologischer Beobachter* 52 (1911): 219–20.

In addition to highlighting the more important contemporary sources used in this study, I would like, finally, to indicate a few particularly important historical studies. On the history of zoological gardens, see, in addition to Knauer's *Der Zoologische Garten*, Gustav Loisel, *Histoire des Menageries* (Paris: O. Doin et fils, 1912); C. V. A. Peel, *The Zoological Gardens of Europe: Their History and Chief Features* (London: F. E. Robinson, 1903); Ellen Velvin, *From Jungle to Zoo* (New York: Moffat, Yard, 1915); Werner Kourist, *400 Jahre Zoo. Im Spiegel der Sammlung Werner Kourist/Bonn* (Cologne: Rheinland-Verlag, 1976); James Fisher, *Zoos of the World* (London: Aldus, 1966); David Hancocks, *Animals and Architecture* (New York: Praeger, 1971), the articles by H. Hediger, C. R. Schmidt, R. E. Honegger, K. Brägger, F. Vogel, and L. Dittrich in the special issue of *Anthos: Landscape Architecture Quarterly* 3 (1971); the articles by G. R. Jones, J. C. Coe, R. Kirchshofer, D. Beisel, M. Bushing, and B. v. Tscharner, in the special issue of *Garten und Landschaft: Journal for Landscape Architecture and Landscape Planning* (January 1985); Sir Peter Chalmers Mitchell, *Centenary History of the Zoological Society of London* (London: Zoological Society of London, 1929); Lord Solly Zuckerman, ed., *The Zoological Society of London, 1826–1976 and Beyond*, Symposium of the Zoological Society of London, Number 40 (London: Zoological Society of London, 1976); Christoph Sherpner, *Von Bürgern für Bürger–125 Jahre Zoologischer Garten Frankfurt a.M.* (Frankfurt: Zoologischer Garten, 1983); the memoirs of Ludwig Heck, longtime director of the Berlin Zoological Gardens, published as *Heiter-ernste Lebensbeichte: Erinnerungen eines alten Tiergärtners* (Berlin: Im Deutschen Verlag, 1938); Heinz-Georg Klös, *Berlin und sein Zoo* (Berlin: Haude und Spener, 1978) and *Von der Menagerie zum Tierparadies: 125 Jahre Zoo Berlin* (Berlin: Haude und Spener, 1969); Johann Jakob Hässlin, *Der Zoologische Garten zu Köln: Ein Beitrag zur Geschichte der Tiergärten* (Cologne: Greven, 1960); Harriet Ritvo, *The Animal Estate: The English and Other Creatures in the Victorian Age* (Cambridge: Harvard University Press, 1987); Bob Mullen and Garry Marvin, *Zoo Culture* (London: George Weidenfeld, 1987); Linda Koebner, *Zoo Book: The Evolution of Wildlife Conservation Centers* (New York: Forge, 1994); R. J. Hoage and William Deiss, ed., *New Worlds, New Animals: From Menagerie to Zoological Park in the Nineteenth Century* (Baltimore: Johns Hopkins University Press, 1996); Jeffrey Hyson, "Jungles of Eden: The Design of American Zoos," in *Environmentalism in Landscape Architecture*, ed. Michel Conan (Washington, D.C.: Dumbarton Oaks, 2000), 23–44; and Vernon Kisling Jr., ed., *Zoo and Aquarium History: Ancient Animal Collections to Zoological Gardens* (Boca Raton, Fla.: CRC, 2001).

On the exhibition of indigenous peoples in Europe in the years before World War I, see in particular Thomas Theye's edited collection *Wir und die Wilden: Einblicke in eine kannibalische Beziehung* (Reinbek bei Hamburg: Rowohlt, 1984); Hermann Pollig and Tilman Osterwold, eds. *Exotische Welten, Europäische Phantasien* (Stuttgart: Edition Cantz, 1987);

Volker Harms, ed., *Andenken an den Kolonialismus. Eine Ausstellung des Völkerkundlichen Instituts der Universität Tübingen* (Tübingen: ATTEMPTO, 1984); Christian F. Feest, ed., *Indians and Europe: An Interdisciplinary Collection of Essays* (Aachen: Herodot, 1987); Urs Bitterli, *Die "Wilden" und die "Zivilisierten": Grundzüge einer Geistes- und Kulturgeschichte der europäisch-überseeischen Begegnung* (Munich: Beck, 1976); and, on Hagenbeck in particular, Hilke Thode-Arora's very important 1989 *Für fünfzig Pfennig um die Welt*. On the life and importance of Rudolf Virchow, see especially Erwin H. Ackerknecht, *Rudolf Virchow: Doctor, Statesman, Anthropologist* (Madison: University of Wisconsin Press, 1953) and the *Virchow-Bibliographie, 1843–1901*, ed. J. Schwalbe (New York: Arno Press, 1981).

For recent historical work on the firm of Carl Hagenbeck, see Lothar Dittrich and Annelore Rieke-Müller's 1998 *Carl Hagenbeck (1844–1913): Tierhandel und Schaustellungen im Deutschen Kaiserreich* and Matthias Gretzchel and Ortwin Pelc's 1998 *Hagenbeck: Tiere, Menschen, Illusionen*, both indicated above, and the seminal articles by Herman Reichenbach, including "Carl Hagenbeck's Tierpark and Modern Zoological Gardens," *Journal of the Society for the Bibliography of Natural History* 9 (1980): 573–85, and "A Tale of Two Zoos: The Hamburg Zoological Garden and Carl Hagenbeck's Tierpark," in Hoage and Deiss, *New Worlds, New Animals*, 51–62. See also Erna Mohr, "Das Geschlecht Hagenbeck," *Der Zoologische Garten* 21, no. 1 (1954): 2–9; Alexander Sokolowsky, "Kleine Hagenbeck-Erinnerungen," *Der Zoologische Garten* 21, no. 1 (1954): 10–11; the impressionistic *Im Paradies der Tiere: Kleine Begebenheiten bei Hagenbeck*, by Ilse Bock (Hamburg: Hammerich and Lesser, n.d. [ca. 1948]); Lorenz Hagenbeck's memoirs published as *Den Tieren gehört mein Herz* (Hamburg: Hoffmann und Campe, 1955); and Günter H. W. Niemeyer, *Hagenbeck: Geschichte und Geschichten* (Hamburg: Hans Christians, 1972).

Index

Casanova, Lorenzo: as animal catcher,
49–50, 52, 53, 55, 57, 59; *vs.* other ani-
mal catchers, 71, 80
Castan's Panoptikum, 107, 133
Ceylon shows: *vs.* Fuegian show, 114; and
John Hagenbeck, 76; qualities of, 85–
86; surplus elephants from, 148. *See also*
Indian shows; Sinhalese shows
Circus Hagenbeck. *See* Hagenbeck's
circuses
Clara (rhinoceros), travels of, 14–15, 208n. 1
Cohn, Bernard: as animal catcher, 51, 53,
55, 57; and Sudan show, 83–84
Columbus, Christopher, and exhibitions
of people, 86–87
Consul (chimpanzee), as civilized ape,
190
Cook, James, 87, 136
Croke, Vicki, on immersion exhibits in
zoological gardens, 200
Crowther, Miss (gorilla). *See* Miss Crowther

Darwin, Charles: on the Fuegians 118, 119;
and Miss Crowther (gorilla), 189; *The
Origin of Species,* and exhibitions of
people, 111
Disney's Animal Kingdom, 43, 197
Dominik, Hans: as animal catcher, 51; on
extinction of elephants, 218n. 54; *vs.*
William Temple Hornaday, 67; *Kamerun: Sechs Kriegs- und Friedensjahre in
deutschen Tropen,* 60–63; military
tactics of, 67–68; *vs.* Hans Hermann
Schomburgk, 65–66
Du Chaillu, Paul, *Explorations and Adventures in Equatorial Africa,* on gorillas,
2–5, 6, 16
Dürer, Albrecht, *vs.* Jean-Baptiste Oudry,
14

Edison, Thomas Alva, 206
Egeh, Hersy, and the Somali shows, 237n.
149

Eggenschwyler, Urs, and Hagenbeck's
Animal Park, 162
Elephants: and Phineas T. Barnum and
Adam Forepaugh, 58, 85; and Berlin
Zoological Gardens, 35–36; and Hans
Dominik, 60–66; in early animal shipments, 49, 50, 54–56, 57–58; and exhi-
bitions of people, 85–86, 106, 129; and
Hagenbeck circuses, 148, 149; for hunt-
ing parks, 162; and mammoths, 110, 120–
25; and Wilhelm Munnecke, 76–80,
193; 1913 shipment of, 187; prices for,
180; and Hans Hermann Schomburgk,
63–67, 218n. 56, 219n. 62, 219–20n. 65;
and Thomas Hosmer Shepherd, 33; at
World's Columbian Exposition, 151–52;
and Zoological Gardens of London's
1965 pavilion, 200
Elven, Eduard, description of Christoph
Schulz, 70
Eskimo shows. *See* Greenland show;
Labrador show
Ethiopian show, 129–30
Ethnology: *vs.* anthropology, 103–5; and
exhibitions of people, 103–10; meaning
of term, 229–30n. 65. *See also* Anthropology; Exhibitions of people;
Prehistory
Eugene of Savoy, Prince: Belvedere
menagerie of, 25–31; as collector, 26–27
Exhibitions of people: and anthropology,
95–105, 228–29n. 56; and authenticity,
88–90, 126–28, 141–45, 182, 194; criticism of, 96–97, 140–41; demise of, 12,
141–45, 194–95, 204; distinctiveness of
Hagenbeck's shows, 86–90; eroticism
of, 131–36; and ethnology, 103–10; and
photography, 100–104, 108–9, 131–36;
and prehistory, 110–25; reasons for, 81–
82, 136–41; and science, 91–125; at
World's Columbian Exposition, 90–91.
See also individual shows

Nansen, Fridtjof, and historical panoramas, 167, 171, 175

National Aquarium (Baltimore, MD), 205–6

National Zoo (Washington, D.C.), 11

Natural history museums, 6–7; *vs.* zoological gardens, 11

"Nature Faker" controversy, 69, 221n. 75

Neanderthal man: and studies of prehistoric Europe, 111; Rudolf Virchow on, 227n. 40

Nelson, Richard, *Heart and Blood: Living with Deer in America,* 11–12

Neschiwow, Osip, and Alexander Sokolowsky, 186–87

New Caledonia show, 133, 141–42

Niemeyer, Günter H. W.: and exhibitions of people, 204; on Greenland show, 139; on modern zoological gardens, 183

Nubian shows, Rudolf Virchow on, 98–99, 100, 133

Oettermann, Stephan, and panoramas, 169–70

Okabak, Caspar Mikel: and Greenland show, 84–85, 93–95; identity of, 125; and Günter H. W. Niemeyer, 139; photograph of, 94

Omai, exhibition of, 87

Ontogeny and phylogeny, theory of, 111–12

Oregon Coast Aquarium, 202

Ostrich farms, 244–45n. 75

Ota Benga, exhibition of, 39

Oudry, Jean-Baptiste, *Study of a Rhinoceros,* 14

Pandas, fascination with, 11, 201

Panoramas: and birth of modern zoo designs, 183–84; of Cameroon, 169; and "Carl Hagenbeck's Zoological Paradise," 165–67, 171; criticism of, 171–75; at Hagenbeck's Animal Park, 163, 168, 171–75; of Fridtjof Nansen's *Fram,* 167, 169, 171, 175; and Anton von Werner, 169–71

People shows. *See* Exhibitions of people

Philadelphia, William, and Hagenbeck's circuses, 157–58

Ploss, Hermann, and Max Bartels, 102–3, 136

Prehistory: and exhibitions of "primitive" people, 110–25; and Sigmund Freud's *Totem and Taboo,* 113–14; and Fuegian show, 114–19; and Heinrich Leutemann, 120–25; and David Friedrich Weinland's *Rulaman,* 110–14, 120–24. *See also* Anthropology; Ethnology; Exhibitions of people

Priemel, Kurt, on Hagenbeck's Animal Park, 172–73, 176, 177, 186

Primate tea parties, and zoological gardens, 191

"Prince Dido of Didotown," 126–27

Przewalski's horse, 52, 161, 180

Pygmy hippopotamus, and Hans Hermann Schomburgk, 58, 68–69, 181, 219n. 61, 221n. 74, 245–46n. 80

Raffles, Sir Thomas Stamford, and Zoological Society of London, 31–32

Ranke, Johannes: and anthropological photographs, 102; on Fuegians and stone age Europeans, 119

Ratzel, Friedrich, and anthropological photographs, 102

Recapitulation debate, 231–32n. 81

Red Peter (civilized ape). *See* Kafka, Franz

Reichenbach, Herman: on Carl Hagenbeck Sr., 213n. 5; on Hagenbeck's animal trade, 181, 214n. 19

Renz, Christian, and traveling menageries, 47

"A Report to an Academy." *See* Kafka, Franz

Reynard the Fox, 189

Reynolds, Sir Joshua, and Omai, 87

Rice, Charles, and Carl Hagenbeck, 213n. 8

Riefenstahl, Leni, and the Nuba and Kau, 142

Rilke, Rainer Maria, "The Panther, *Jardin des Plantes, Paris*" (1907), 146, 193–94

Ritvo, Harriet, on history of zoological gardens and empire, 21–22, 209n. 17

Roosevelt, Theodore: and *Hunting Trips of a Ranchman: Sketches of Sport on the Northern Cattle Plains,* 5–6; on Carl G. Schillings, 51, 216n. 27

Rothschild, Walter von, 161, 206

Rulaman. Naturgeschichtliche Erzählung aus der Zeit des Höhlenmenschen und des Höhlenbären (Weinland), 110–14, 120–24

Sami shows, 82–83, 85, 86, 114, 133; authenticity of, 88–89; and breast-feeding, 134; ethnographic materials in, 106; participants in, 225n. 30; photographs of, 103, 105

Samoa show, and eroticism, 134–36, 227n. 42

San Diego Wild Animal Park, 197

Savery, Roelandt, and the dodo, 39

Schaller, George B., on pandas, 212n. 47

Schama, Simon, on European wisent, 24

Schillings, Carl G., as animal catcher, 51, 52, 69, 216n. 27

Schomburgk, Hans Hermann: background of, 51; on catching elephants, 64–65, 219n. 62; on catching pygmy hippopotamuses, 58, 68–69, 181, 219n. 61, 221n. 74, 245–46n. 80; on danger of elephant hunting, 219–20n. 65; *vs.* Hans Dominik and William Temple Hornaday, 65–67; *vs.* other animal catchers, 71, 80; on sentiment in nature, 219n. 64; and *Wild und Wilde im Herzen Afrikas,* 63–67, 218n. 56

Schulz, Christoph: and *Auf Großtierfang für Hagenbeck: Selbsterlebtes aus afrikanischer Wildnis,* 70–73; background of, 51–52; on Hagenbeck's Animal Park, 182; as letter writer, 181; *vs.* other animal catchers, 71, 80; and Hans Hermann Schomburgk, 65; and Alexander Sokolowsky, 186

Schulz, Elisabeth, as animal catcher, 51, 71–72

Sea World, 43, 205

Shasta (liger), and Hogle Zoo (Salt Lake City, Utah), 7–8

Shepherd, Thomas Hosmer, and Zoological Gardens of London, 32–33, 34, 35

Sinhalese shows, 86, 91, 148. *See also* Ceylon shows; Indian shows

Sokolowsky, Alexander: and anthropological and ethnological photographs, 103, 109; *Beobachtungen über die Psyche der Menschenaffen,* 1–2, 3–5, 6, 16; on colonial movement, 136–37; on Carl Hagenbeck's achievements, 195; on history of exhibiting people, 86; on panoramas, 174–75, 241n. 60; on science and the exhibitions of people, 96, 136–37, 225–26n. 38; on sport hunting, 185; on volume of the Hagenbeck animal business, 186–87

Somali shows, 91, 114, 137, 220n. 72

Steiff, stuffed animals of, 69

Steinitz, Heinrich, on the Fuegians, 119

Stellingen. *See* Hagenbeck's Animal Park

Stiansen, Endre, on the gum arabic trade, 56

Stratz, Carl: and anthropological photographs, 101–3; on evolutionary theory, 111–12

Sudan shows, 83–84

Suez Canal, and the animal trade, 49

Terra del Fuego show. *See* Fuegian show

Theye, Thomas, on exhibitions of people, 135–36

Thomas, Keith, on history of zoological gardens, 21

Tierhandel. *See* Hagenbeck's animal trade

Tierpark, Hagenbeck's. *See* Hagenbeck's Animal Park

Triest (lion), and Carl Hagenbeck, 73, 176, 206

Tudge, Colin, on history of zoological gardens, 24

Ukubak. *See* Okabak, Caspar Mikel

Umlauff, Heinrich: and Amazon corps, 133–34; stuffed animals of, 236n. 131

Van Amburgh, Isaac A.: and Sir Edwin Landseer, 158–59; as lion tamer, 158–60

Velvin, Ellen, on gorillas, 207n. 4

Verisimilitude: in Hagenbeck's animal exhibits, 182, 194, 201; in Hagenbeck's exhibitions of people, 89–92, 126, 144–45, 194, 225n. 35, 225n. 37

Victoria, Queen, and Isaac A. Van Amburgh, 158–59

Virchow, Rudolf: on Aztec shows, 98, 226n. 39; on exhibitions of people, 93, 96–98, 125; on Fuegian show, 117–18; on Greenland show, 92–96; importance of, 226–27n. 40; and Labrador Inuit show, 229n. 57; and theory of natural selection, 231n. 80

Völkerausstellungen. *See* Exhibitions of people

Völkerschauen. *See* Exhibitions of people

Vosseler, J., and zoological gardens, 17–18

Walruses, 51, 75–76, 184–85, 222n. 85

Weinland, David Friedrich: on history of zoological gardens, 18; *Rulaman*, 110–14, 120–24

Werner, Anton von, and battle of Sedan, 169–71

Wilhelm II, Kaiser: and exhibitions of people, 137–38; and Carl Hagenbeck, 206

World's Columbian Exposition of 1893: exhibitions of people at, 90–91; and Carl Hagenbeck, 149–55; and Josef Menges, 106

Zoological gardens: and animal trade, 69–70, 177–81, 186–87; and aquariums, 204–6; disappointment of, 10–11; *vs.* Hagenbeck's Animal Park, 163–64, 171–75, 183–84, 199; historiography of, 16–25; history of, 13–14; immersion exhibits in, 200–202; *vs.* menageries, 22–23, 37–41; and "silencing" of animals, 12, 197–98; sociability of, 33–36; theatrical qualities of, 43; twentieth-century architecture in, 199–200. *See also* Belvedere menagerie; Berlin Zoological Gardens; Hagenbeck's Animal Park; Menageries; Zoological Gardens of London

Zoological Gardens of London: *vs.* Belvedere menagerie, 25, 33–34, 37–38, 43; history of, 13, 19, 25, 31–37, 200

Zuckmayer, Carl, on "the idea" of Carl Hagenbeck, 45

Zukowsky, Ludwig: on capturing walruses, 184–85, 222n. 85; on the goals of Carl Hagenbeck, 42, 175–76; on Hagenbeck's arctic panorama, 173–74; on science and the exhibitions of people, 136; on Triest (lion), 73